A WEEK AT
THE AIRPORT

A HEATHROW DIARY

A WEEK AT THE AIRPORT
A HEATHROW DIARY

ALAIN DE BOTTON
PHOTOGRAPHS BY RICHARD BAKER

P

First published in Great Britain in 2009 by
PROFILE BOOKS LTD
3a Exmouth House
Pine Street
London EC1R 0JH
www.profilebooks.com

Copyright © Alain de Botton, 2009

10 9 8 7 6 5 4 3 2 1

Photographs © Richard Baker, 2009
Text designed and typeset in ITC Galliard
by Joana Niemeyer at April
Cover design by David Pearson
Printed and bound in Great Britain
by Butler, Tanner and Dennis Ltd, Frome, Somerset

A CIP catalogue record for this book is available from
the British Library.

ISBN 978 1 84668 359 6

Mixed Sources
Product group from well-managed
forests and other controlled sources
www.fsc.org Cert no. SGS-COC-005091
© 1996 Forest Stewardship Council
FSC

For Saul

Contents

I   Approach

1   While punctuality lies at the heart of what we typically understand by a good trip, I have often longed for my plane to be delayed – so that I might be forced to spend a bit more time at the airport. I have rarely shared this aspiration with other people, but in private I have hoped for a hydraulic leak from the undercarriage or a tempest off the Bay of Biscay, a bank of fog in Malpensa or a wildcat strike in the control tower in Málaga (famed in the industry as much for its hot-headed labour relations as for its even-handed command of much of western Mediterranean airspace). On occasion, I have even wished for a delay so severe that I would be offered a meal voucher or, more dramatically, a night at an airline's expense in a giant concrete Kleenex box with unopenable windows, corridors decorated with nostalgic images of propeller planes and foam pillows infused with the distant smells of kerosene.

In the summer of 2009, I received a call from a man who worked for a company that owned airports. It held the keys to Southampton, Aberdeen, Heathrow and Naples, and oversaw the retail operations at Boston Logan and Pittsburgh International. The corporation additionally controlled large pieces of the industrial infrastructure upon which European civilisation relies (yet which we as individuals seldom trouble ourselves about as we use the bathroom in Białystok or drive our rental car to Cádiz): the waste company Cespa, the Polish construction group Budimex and the Spanish toll-road concern Autopista.

My caller explained that his company had lately developed an interest in literature and had taken a decision to invite a writer to spend a week at its newest passenger hub, Terminal 5, situated between the two runways of London's largest airport. This artist, who was sonorously to be referred to as Heathrow's first writer-in-residence, would be asked to conduct an impressionistic survey of the premises and then, in full view of passengers and staff, draw together material for a book at a specially positioned desk in the departures hall between zones D and E.

It seemed astonishing and touching that in our distracted age, literature could have retained sufficient prestige to inspire a multinational enterprise, otherwise focused on the management of landing fees and effluents, to underwrite a venture invested with such elevated artistic ambitions. Nevertheless, as the man from the airport company put it to me over the telephone, with a lyricism as vague as it was beguiling, there were still many aspects of the world that perhaps only writers could be counted on to find the right words to express. A glossy marketing brochure, while in certain contexts a supremely effective instrument of communication, might not always convey the authenticity achievable by a single authorial voice – or, as my friend suggested with greater concision, could more easily be dismissed as 'bullshit'.

2 Though the worlds of commerce and art have frequently been unhappy bedfellows, each viewing the other with a mixture of paranoia and contempt, I felt it would be churlish of me to decline to investigate my caller's offer simply because his company administered airside food courts and hosted technologies likely to be involved in raising the planet's median air temperature. There were undoubtedly some skeletons in the airport company's closet, arising from its intermittent desire to pour cement over age-old villages and its skill in encouraging us to circumnavigate the globe on unnecessary journeys, laden with bags of Johnnie Walker and toy bears dressed up as guards of the British monarchy.

But with my own closet not entirely skeleton-free, I was in no position to judge. I understood that money accumulated on the battlefield or in the marketplace could fairly be redirected towards higher aesthetic ends. I thought of impatient ancient Greek statesmen who had once spent their war spoils building temples to Athena and ruthless Renaissance noblemen who had blithely commissioned delicate frescoes in honour of spring.

Besides, and more prosaically, technological changes seemed to be drawing a curtain on a long and blessed interlude in which writers had been able to survive by selling their works to a wider public, threatening a renewed condition of anxious dependence on the largesse of individual sponsors. Contemplating what it might mean to be employed by an airport, I looked with plaintive optimism to the example of the seventeenth-century philosopher Thomas Hobbes, who had thought nothing of writing his books while in the pay of the Earls of Devonshire, routinely placing florid declarations to them in his treatises and even accepting their gift of a small bedroom next to the vestibule of their home in Derbyshire, Hardwick Hall. 'I humbly offer my book to your Lordship,' England's subtlest political theorist had written to the swaggering William Devonshire on presenting him with *De Cive* in 1642. 'May God of Heaven crown you with many days in your earthly station, and many more in heavenly Jerusalem.'

In contrast, my own patron, Colin Matthews, the chief executive of BAA, the owner of Heathrow, was the most undemanding of employers. He made no requests whatever of me, not for a dedication, or even a modest reference to his prospects in the next world. His staff went so far as to give me explicit permission to be rude about the airport's activities. In such lack of constraints, I felt myself to be benefiting from a tradition wherein the wealthy merchant enters into a relationship with an artist fully prepared for him to behave like an outlaw; he does not expect good manners, he knows and is half delighted by the idea that the favoured baboon will smash his crockery. In such tolerance lies the ultimate proof of his power.

3 In any event, my new employer was legitimately proud of his terminal and understandably keen to find ways to sing of its beauty. The undulating glass and steel structure was the largest building in the land, forty metres tall and 400 long, the size of four football pitches, and yet the whole conveyed a sense of continuous lightness and ease, like an intelligent mind engaging effortlessly with complexity. The blinking of its ruby lights could be seen at dusk from Windsor Castle, the terminal's forms giving shape to the promises of modernity.

Standing before costly objects of technological beauty, we may be tempted to reject the possibility of awe, for fear that we could grow stupid through admiration. We may feel at risk of becoming overimpressed by architecture and engineering, of being dumbstruck by the Bombardier trains that progress driverlessly between satellites or by the General Electric GE90 engines that hang lightly off the composite wings of a Boeing 777 bound for Seoul.

And yet to refuse to be awed at all might in the end be merely another kind of foolishness. In a world full of chaos and irregularity, the terminal seemed a worthy and intriguing refuge of elegance and logic. It was the imaginative centre of contemporary culture. Had one been asked to take a Martian to visit a single place that neatly captures the gamut of themes running through our civilisation – from our faith in technology to our destruction of nature, from our interconnectedness to our romanticising of travel – then it would have to be to the departures and arrivals halls that one would head. I ran out of reasons not to accept the airport's unusual offer to spend a little more time on its premises.

II   Departures

I arrived at the airport on a train from central London early on a Sunday evening, a small roller case in hand and no further destination for the week. I had been billeted at the Terminal 5 outpost of the Sofitel hotel chain, which, while not directly under the ownership of the airport, was situated only a few metres away from it, umbilically connected to the mothership by a sequence of covered walkways and a common architectural language featuring the repeated use of glazed surfaces, giant potted vegetation and grey tiling.

The hotel boasted 605 rooms that faced one another across an internal atrium, but it soon became evident that the true soul of the enterprise lay not so much in hostelry as in the management of a continuous run of conferences and congresses, held in forty-five meeting rooms, each one named after a different part of the world, and well equipped with data points and LAN facilities. At the end of this August Sunday, Avis Europe was in the Dubai Room and Liftex, the association of the British lift industry, in the Tokyo Hall. But the largest gathering was in the Athens Theatre, where delegates were winding up a meeting about valve sizes chaired by the International Organization for Standardization (or ISO), a body committed to eradicating incompatibilities between varieties of industrial equipment. So long as the Libyan government honoured its agreements, thanks to twenty years of work by the ISO, one would soon be able to travel across North Africa, from Agadir to El Gouna, without recourse to an adapter plug.

2   I had been assigned a room at the top western corner of the building, from which I could see the side of the terminal and a sequence of red and white lights that marked the end of the northern runway. Every minute, despite the best attempts of the glazing contractors, I heard the roar of an ascending jet, as hundreds of passengers, some perhaps holding their partners' hands, others sanguinely scanning *The Economist*, submitted themselves to a calculated defiance of our species' land-based origins. Behind each successful flight lay the coordinated efforts of hundreds of souls, from the manufacturers of airline amenity kits to the Honeywell engineers responsible for installing windshear-detection radars and collision-avoidance systems.

The hotel room appeared to have taken its design cues from the business-class cabin – though it was hard to say for sure which had inspired which, whether the room was skilfully endeavouring to look like a cabin, or the cabin a room – or whether they simply both shared in an unconscious spirit of their age, of the kind that had once ensured continuity between the lace trim on mid-eighteenth-century evening dresses and the iron detailing on the façades of Georgian town houses. The space held out the promise that its occupant might summon up a film on the adjustable screen, fall asleep to the drone of the air-conditioning unit and wake up on the final descent to Chek Lap Kok.

My employer had ordered me to remain within the larger perimeter of the airport for the duration of my seven-day stay and had accordingly provided me with a selection of vouchers from the terminal's restaurants as well as authorisation to order two evening meals from the hotel.

There can be few literary works in any language as poetic as a room-service menu.

> The autumn blast
> Blows along the stones
> On Mount Asama

Even these lines by Matsuo Bashō, who brought the haiku form to its mature perfection in the Edo era in Japan, seemed flat and unevocative next to the verse composed by the anonymous master at work somewhere within the Sofitel's catering operation:

> Delicate field greens with sun-dried cranberries,
> Poached pears, Gorgonzola cheese
> And candied walnuts in a Zinfandel vinaigrette

I reflected on the difficulty faced by the kitchen of correctly interpreting the likelihood of selling some of the remoter items of the menu: how many out of the guests in the lift industry, for example, might be tempted by the 'Atlantic snapper, enhanced with lemon pepper seasoning atop a chunky mango relish', or by the always mysterious and somewhat melancholy-sounding 'Chef's soup of the day'. But perhaps, in the end, there was no particular science to the calibration of alimentary supplies, for it is rare to spend an evening in a hotel and order anything other than a club sandwich, which even Bashō, at the peak of his powers, would have struggled to describe as convincingly as the menu's scribe:

Warm grilled chicken slices,
Smoked bacon, crisp lettuce,
And a warm ciabatta roll on a bed of sea-salted fries

There was a knock at the door only twenty minutes after I had dialled nine and put in my order. It is a strange moment when two adult men meet each other, one naked save for a complimentary dressing gown, the other (newly arrived in England from the small Estonian town of Rakvere and sharing a room with four others in nearby Hillingdon) sporting a black and white uniform, with an apron and a name badge. It is difficult to think of the ritual as entirely unremarkable, to say in a casually impatient voice, 'By the television, please,' while pretending to rearrange papers – though this capacity can be counted upon to evolve with more frequent attendance at global conferences.

I had dinner with Chloe Cho, formerly with Channel NewsAsia but now working for CNBC in Singapore. She updated me on the regional markets and Samsung's quarterly forecast, but her sustained focus was on commodities. I wondered what Chloe's outside interests might be. She was like a sister of the Carmelite Order, behind whose austere headdress and concentrated expression one could just guess at occasional moments of doubt, rendered all the more intriguing by their emphatic denial. On a ticker tape running across the bottom of the screen, I spotted the share price of my employer, pointed on a downward trajectory.

After dinner, it was still warm and not yet quite dark outside. I would have liked to take a walk around one of the few fields that remained of the farmland on which the airport had been built some six decades before, but it seemed at once perilous and impossible to leave the building, so I decided to do a few circuits around the hotel corridors instead. Feeling disoriented and queasy, as if I were on a cruise ship in a swell, I repeatedly had to steady myself against the synthetic walls. Along my route, I passed dozens of room-service trays much like my own, each one furtively pushed into the hallway and nearly all (once their stainless-steel covers were lifted) providing evidence of orgiastic episodes of consumption. Ketchup smeared across slices of toast and fried eggs dipped in vinaigrette spoke of the breaking of taboos just like the sexual ones more often assumed to be breached during solitary residence in hotel rooms.

I fell asleep at eleven, but woke up again abruptly just past three. The prehistoric part of the mind, trained to listen for and interpret every shriek in the trees, was still doing its work, latching on to the slamming of doors and the flushing of toilets in unknown precincts of the building. The hotel and terminal seemed like a giant machine poised in standby mode, emitting an uncanny hum from a phalanx of slowly rotating exhaust fans. I thought of the hotel's spa, its hot tubs perhaps still bubbling in the darkness. The sky was a chemical orange colour, observing the final hours of the fragile curfew it had been keeping ever since it had swallowed up the last of the previous evening's Asia-bound flights. Jutting from the side of the terminal was the disembodied tail of a British Airways A321, anticipating another imminent odyssey in the merciless cold of the lower stratosphere.

3 In the end it was a 5.30 a.m. arrival (BA flight from Hong Kong) that called a halt to my perturbed night. I showered, ate a fruit bar purchased from a dispensing machine in the car park and wandered over to an observation area next to the terminal. In the cloudless dawn, a sequence of planes, each visible as a single diamond, were lined up at different heights, like pupils in a school photo, on their final approach to the northern runway. Their wings unfolded themselves into elaborate and unlikely

arrangements of irregularly sized steel-grey panels. Having avoided the earth for so long, wheels that had last touched ground in San Francisco or Mumbai hesitated and slowed almost to a standstill as they arched and prepared to greet the rubber-stained English tarmac with a burst of smoke that made manifest their planes' speed and weight.

With the aggressive whistling of their engines, the airborne visitors appeared to be rebuking this domestic English morning for its somnolence, like a delivery person unable to resist pressing a little too insistently and vengefully on the doorbell of a still-slumbering household. All around them, the M4 corridor was waking up reluctantly. Kettles were being switched on in Reading, shirts being ironed in Slough and children unfurling themselves beneath their Thomas the Tank Engine duvets in Staines.

Yet for the passengers in the 747 now nearing the airfield, the day was already well advanced. Many would have awakened several hours before to see their plane crossing over Thurso at the northernmost tip of Scotland, nearly the end of the earth to those in London's suburbs, but their destination's very doorstep for travellers after a long night's journey over the Canadian icelands and a moonlit North Pole. Breakfast would have kept time with the airliner's progress down the spine of the kingdom: a struggle with a small box of cornflakes over Edinburgh, an omelette studded with red peppers and mushrooms near Newcastle, a stab at a peculiar-looking fruit yoghurt over the unknowing Yorkshire Dales.

For British Airways planes, the approach to Terminal 5 was a return to their home base, equivalent to the final run up the Plymouth Sound for their eighteenth-century naval predecessors. Having long been guests on foreign aprons, allotted awkward and remote slots at O'Hare or LAX, the odd ones out amid immodestly long rows of United and Delta aircraft, they now took their turn at having the superiority of numbers, lining up in perfect symmetry along the back of Satellite B.

Sibling 747s that had only recently been separated out across the world were here parked wing tip to wing tip, Johannesburg next to Delhi, Sydney next to Phoenix. Repetition lent their fuselage designs a new beauty: the eye could follow a series of identical motifs down a fifteen-strong line of dolphin-like bodies, the resulting aesthetic effect only enhanced by the knowledge that each plane had cost some $250 million, and that what lay before one was therefore a symbol not just of the modern era's daunting technical intelligence but also of its prodigious and inconceivable wealth.

As every plane took up its position at its assigned gate, a choreographed dance began. A passenger walkway rolled forward and closed its rubber mouth in a hesitant kiss over the front left-hand door. A member of the ground staff tapped at the window, a colleague inside released the airlock and the two airline personnel exchanged the sort of casual greeting one might have expected between office workers returning to adjacent desks after lunch, rather than the encomium that would more fittingly have marked the end of an 11,000-kilometre journey from the other side of the globe. Then again, the welcome may be no more effusive a hundred years hence, when, at the close of a nine-month voyage, against the eerie blood-red midday light bathing a spaceport in Mars's Cydonian hills, a fellow human knocks at the gold-tinted window of our just-docked craft.

Cargo handlers opened the holds to unload crates filled with the chilled flanks of Argentine cattle and the crenellated forms of crustaceans that had, just the day before, been marching heedlessly across Nantucket Sound. In only a few hours, the plane would be sent up into the sky once more. Fuel hoses were attached to its wings and the tanks replenished with Jet A-1 that would steadily be burned over the African savannah. In the already vacant front cabins, where it might cost the equivalent of a small car to spend the night reclining in an armchair, cleaners scrambled to pick up the financial weeklies, half-eaten chocolates and distorted foam earplugs left behind by the flight's complement of plutocrats and actors. Passengers disembarked for whom this ordinary English morning would have a supernatural tinge.

4 Meanwhile, at the drop-off point in front of the terminal, cars were pulling up in increasing numbers, rusty minicabs with tensely negotiated fares alongside muscular limousines from whose armoured doors men emerged crossly and swiftly into the executive channels.

Some of the trips starting here had been decided upon only in the previous few days, booked in response to a swiftly developing situation in the Munich or Milan office; others were the fruit of three years' painful anticipation of a return to a village in northern Kashmir, with six dark-green suitcases filled with gifts for young relatives never previously met.

The wealthy tended to carry the least luggage, for their rank and itineraries led them to subscribe to the much-published axiom that one can now buy anything anywhere. But they had perhaps never visited a television retailer in Accra or they might have looked more favourably upon a Ghanaian family's decision to import a Samsung PS50, a high-definition plasma machine the weight and size of a laden coffin. It had been acquired the day before at a branch of Comet in Harlow and was eagerly awaited in the Kissehman quarter of Accra, where its existence would stand as evidence of the extraordinary status of its importer, a thirty-eight-year-old dispatch driver from Epping.

Entry into the vast space of the departures hall heralded the opportunity, characteristic in the transport nodes of the modern world, to observe people with discretion, to forget oneself in a sea of otherness and to let the imagination loose on the limitless supply of fragmentary stories provided by the eye and ear. The mighty steel bracing of the airport's ceiling recalled the scaffolding of the great nineteenth-century railway stations, and evoked the sense of awe – suggested in paintings such as Monet's *Gare Saint-Lazare* – that must have been experienced by the first crowds to step inside these light-filled, iron-limbed halls pullulating with strangers, buildings that enabled a person to sense viscerally, rather than just grasp intellectually, the vastness and diversity of humanity.

The roof of the building weighed 18,000 tonnes, but the steel columns supporting it hardly suggested the pressures they were under. They were endowed with a subcategory of beauty we might refer to as elegance, present whenever architecture has the modesty not to draw attention to the difficulties it has surmounted. On top of their tapered necks, the columns balanced the 400-metre roof as if they were holding up a canopy made of linen, offering a metaphor for how we too might like to stand in relation to our burdens.

Most passengers were bound for a bank of automatic check-in machines in the centre of the hall. These represented an epochal shift away from the human hand and towards the robot, a transition as significant in the context of airline logistics as that from the washboard to the washing machine had once been in the domestic sphere. However, few users seemed capable of producing the precise line-up of cards and codes demanded by the computers, which responded to the slightest infraction with sudden and intemperate error messages – making one long for a return of the surliest of humans, from whom there always remains at least a theoretical possibility of understanding and forgiveness.

Nowhere was the airport's charm more concentrated than on the screens placed at intervals across the terminal which announced, in deliberately workmanlike fonts, the itineraries of aircraft about to take to the skies. These screens implied a feeling of infinite and immediate possibility: they suggested the ease with which we might impulsively approach a ticket desk and, within a few hours, embark for a country where the call to prayer rang out over shuttered whitewashed houses, where we understood nothing of the language and where no one knew our identities. The lack of detail about the destinations served only to stir unfocused images of nostalgia and longing: Tel Aviv, Tripoli, St Petersburg, Miami, Muscat via Abu Dhabi, Algiers, Grand Cayman via Nassau... all of these promises of alternative lives, to which we might appeal at moments of claustrophobia and stagnation.

5 A few zones of the check-in area remained dedicated to traditionally staffed desks, where passengers were from the start assured of interaction with a living being. The quality of this

interaction was the responsibility of Diane Neville, who had worked for British Airways since leaving school fifteen years before and now oversaw a staff of some two hundred who dispensed boarding cards and affixed luggage labels.

It was never far from Diane's thoughts how vulnerable her airline was to its employees' bad moods. On reaching home, a passenger would remember nothing of the plane that had not crashed or the suitcase that had arrived within minutes of the carousel's starting if, upon politely asking for a window seat, she had been brusquely admonished to be happy with whatever she was assigned – this retort stemming from a sense on the part of a member of the check-in team (perhaps discouraged by a bad head cold or a disappointing evening at a nightclub) of the humiliating and unjust nature of existence.

In the earliest days of industry, it had been an easy enough matter to motivate a workforce, requiring only a single and basic tool: the whip. Workers could be struck hard and with impunity to encourage them to quarry stones or pull on their oars with greater enthusiasm. But the rules had had to be revised with the development of jobs – by the early twenty-first century comprising the dominant sector of the market – that could be successfully performed only if their protagonists were to a significant degree satisfied rather than resentfully obedient. Once it became evident that someone who was expected to wheel elderly passengers around a terminal, for example, or to serve meals at high altitudes could not profitably be sullen or furious, the mental well-being of employees began to be a supreme object of commercial concern.

Out of such requirements had been born the art of management, a set of practices designed to coax rather than simply extort commitment out of workers, and which, at British Airways, had inspired the use of regular motivational training seminars, gym access and free cafeterias in order to achieve that most calculated, unsentimental and fragile of goals: a friendly manner.

But however skilfully designed its incentive structure, the airline could in the end do very little to guarantee that its staff would actually add to their dealings with customers that almost imperceptible measure of goodwill which elevates service from mere efficiency to tangible warmth. Though one can inculcate competence, it is impossible to legislate for humanity. In other words, the airline's survival depended upon qualities that the company itself could not produce or control, and was not even, strictly speaking, paying for. The real origins of these qualities lay not in training courses or employee benefits but, for example, in the loving atmosphere that had reigned a quarter of a century earlier in a house in Cheshire, where two parents had brought up a future staff member with benevolence and humour – all so that today, without any thanks being given to those parents (a category deserving to be generally known as the true Human Resources department of global capitalism), he would have both the will and the wherewithal to reassure an anxious student on her way to the gate to catch BA048 to Philadelphia.

6 But even true friendliness was not always enough. I observed a passenger running with shoulder bags towards a check-in desk for a Tokyo flight, only to be courteously informed that he had arrived too late to board and would have to consider alternatives.

Yet his 747 had not already departed – it would sit at the terminal for a further twenty minutes, its fuselage visible through the windows. The problem was a purely administrative one: the airline had stipulated that no passenger, even one awaited by a bride and two hundred guests, could be issued with a boarding card less than forty minutes before departure.

The presence of the aircraft combined with its unreachability, the absence of another seat on a flight for forty-eight hours, the cancellation of a day of meetings in Tokyo, all these pushed the man to bang his fists on the counter and let out a scream so powerful that it could be heard as far away as the WH Smith outlet at the western end of the terminal.

I was reminded of the Roman philosopher Seneca's treatise *On Anger*, written for the benefit of the Emperor Nero, and in particular of its thesis that the root cause of anger is hope. We are angry because we are overly optimistic, insufficiently prepared for the frustrations endemic to existence. A man who screams every time he loses his keys or is turned away at an airport is evincing a touching but recklessly naïve belief in a world in which keys never go astray and our travel plans are invariably assured.

Given Seneca's analysis, it was ominous to note the direction that the airline was taking in its advertising. It was promising ever more confidently to try its very best to serve, to please and to be punctual. As a result, in an industry as vulnerable to disaster as this one, there were surely many more screams to come.

7   Not far from the incautiously hopeful man, a pair of lovers were parting. She must have been twenty-three, he a few years older. There was a copy of Haruki Murakami's *Norwegian Wood* in her bag. They both wore oversize sunglasses and had come of age in the period between SARS and swine flu. It was the intensity of their kiss that first attracted my attention, but what had seemed like passion from afar was revealed at closer range to be an unusual degree of devastation. She was shaking with sorrowful disbelief as he cradled her in his arms and stroked her wavy black hair, in which a clip shaped like a tulip had been fastened. Again and again, they looked into each other's eyes and every time, as though made newly aware of the catastrophe about to befall them, they would begin weeping once more.

Passers-by evinced sympathy. It helped that the woman was extraordinarily beautiful. I missed her already. Her beauty would have been an important part of her identity from at least the age of twelve and, in its honour, she would occasionally pause and briefly consider the effect of her condition on her audience before returning to her lover's chest, damp with her tears.

We might have been ready to offer sympathy, but in actuality there were stronger reasons to want to congratulate her for

having such a powerful motive to feel sad. We should have envied her for having located someone without whom she so firmly felt she could not survive, beyond the gate let alone in a bare student bedroom in a suburb of Rio. If she had been able to view her situation from a sufficient distance, she might have been able to recognise this as one of the high points in her life.

There seemed no end to the ritual. The pair would come close to the security zone, then break down again and retreat for another walk around the terminal. At one point, they went down to the arrivals hall and for a moment it looked as if they might go outside and join the queue at the taxi rank, but they were only buying a packet of dried mango slices from Marks and Spencer, which they fed to each other with pastoral innocence. Then quite suddenly, in the middle of an embrace by the Travelex desk, the beauty glanced down at her watch and, with all the self-control of Odysseus denying the Sirens, ran away from her tormentor down a corridor and into the security zone.

My photographer and I divided forces. I followed her airside and watched her remain stoic until she reached the concourse, only to founder again at the window of Kurt Geiger. I finally lost her in a crowd of French exchange students near Sunglass Hut. For his part, Richard pursued the man down to the train station, where the object of adoration boarded the express service for central London, claimed a seat and sat impassively staring out the window, betraying no sign of emotion save for an unusual juddering movement of his left leg.

8 For many passengers, the terminal was the starting point of short-haul business trips around Europe. They might have announced to their colleagues a few weeks before that they would be missing a few days in the office to fly to Rome, studiously feigning weariness at the prospect of making a journey to the wellspring of European culture – albeit to its frayed edges in a business park near Fiumicino airport.

They would think of these colleagues as they crossed over the Matterhorn, its peak snow-capped even in summer. Just as breakfast was being served in the cabin, their co-workers would be coming into the office – Megan with her carefully prepared lunch, Geoff with his varied ring tones, Simi with her permanent frown – and all the while the travellers would be witnessing below them the byproducts of the titanic energies released by the collision of the Eurasian and African continental plates during the late Mesozoic era.

What a relief it would be for the travellers not to have time to see anything at all of Rome's history or art. And yet how much they would notice nevertheless: the fascinating roadside advertisements for fruit juice on the way from the airport, the unusually delicate shoes worn by Italian men, the odd inflections in their hosts' broken English. What interesting new thoughts would occur to them in the Novotel, what inappropriate films they would watch late into the night and how heartily they would agree, upon their return, with the truism that the best way to see a foreign country is to go and work there.

9 A full 70 per cent of the airport's departing passengers were off on trips for pleasure. It was easy to spot them at this time of year, in their shorts and hats. David was a thirty-eight-year-old shipping broker, and his wife, Louise, a thirty-five-year-old full-time mother and ex-television producer. They lived in Barnes with their two children, Ben, aged three, and Millie, aged five. I found them towards the back of a check-in line for a four-hour flight to Athens. Their final destination was a villa with a pool at the Katafigi Bay resort, a fifty-minute drive away from the Greek capital in a Europcar Category C vehicle.

It would be difficult to overestimate how much time David had spent thinking about his holiday since he had first booked it, the previous January. He had checked the weather reports online every day. He had placed the link to the Dimitra Residence in his Favourites folder and regularly navigated to it, bringing up images of the limestone master bathroom and of the house at dusk, lit up against the rocky Mediterranean slopes. He had pictured himself playing with the children in the palm-lined garden and eating grilled fish and olives with Louise on the terrace.

But although David had reflected at length on his stay in the Peloponnese, there were still many things that managed to surprise him at Terminal 5. He had omitted to recall the existence of the check-in line or to think of just how many people can be fitted into an Airbus A320. He had not focused on how long four hours can seem nor had he considered the improbability of all the members of a family achieving physical and psychological satisfaction at approximately the same time. He had not remembered how hurtful he always found it when Ben made it clear that he disproportionately favoured his mother or how he himself invariably responded to such rejections by becoming unproductively strict, which in turn upset his wife, who liked

to voice her opinion that Ben's reticence was due primarily to the lack of paternal contact he had had since his father's promotion. David's work was a continuous flash-point in the couple's relationship and had in fact precipitated an argument only the night before, during which David had described Louise as ungrateful for failing to appreciate and honour the necessary connection between his absences and their affluence.

Had the plane on which they were to fly to Athens burst into flames shortly after take-off and begun plunging towards the Staines reservoir, David would have clasped the members of his family tightly to him and told them with wholehearted sincerity that he loved them unreservedly – but right now, he could not look a single one of them in the eye.

It seems that most of us could benefit from a brush with a near-fatal disaster to help us to recognise the important things that we are too defeated or embittered to recognise from day to day.

As David lifted a suitcase on to the conveyor belt, he came to an unexpected and troubling realisation: that he was bringing *himself* with him on his holiday. Whatever the qualities of the Dimitra Residence, they were going to be critically undermined by the fact that *he* would be in the villa as well. He had booked the trip in the expectation of being able to enjoy his children, his wife, the Mediterranean, some spanakopita and the Attic skies, but it was evident that he would be forced to apprehend all of these through the distorting filter of his own being, with its debilitating levels of fear, anxiety and wayward desire.

There was, of course, no official recourse available to him, whether for assistance or complaint. British Airways did, it was true, maintain a desk manned by some unusually personable employees and adorned with the message: 'We are here to help'. But the staff shied away from existential issues, seeming to restrict their insights to matters relating to the transit time to adjacent satellites and the location of the nearest toilets.

Yet it was more than a little disingenuous for the airline to deny all knowledge of, and responsibility for, the metaphysical well-being of its customers. Like its many competitors, British Airways, with its fifty-five Boeing 747s and its thirty-seven Airbus A320s, existed in large part to encourage and enable people to go and sit in deckchairs and take up (and usually fail at) the momentous challenge of being content for a few days. The tense atmosphere now prevailing within David's family was a reminder of the rigid, unforgiving logic to which human moods are subject, and which we ignore at our peril when we see a picture of a beautiful house in a foreign country and imagine that happiness must inevitably accompany such magnificence. Our capacity to derive pleasure from aesthetic or material goods seems critically dependent on our first satisfying a more important range of emotional and psychological needs, among them those for understanding, compassion and respect. We cannot enjoy palm trees and azure pools if a relationship to which we are committed has abruptly revealed itself to be suffused with incomprehension and resentment.

There is a painful contrast between the enormous objective projects that we set in train, at incalculable financial and environmental cost – the construction of terminals, of runways and of wide-bodied aircraft – and the subjective psychological knots that undermine their use. How quickly all the advantages of technological civilisation are wiped out by a domestic squabble. At the beginning of human history, as we struggled to light fires and to chisel fallen trees into rudimentary canoes, who could have predicted that long after we had managed to send men to the moon and aeroplanes to Australasia, we would still have such trouble knowing how to tolerate ourselves, forgive our loved ones and apologise for our tantrums?

10 My employer had made good on the promise of a proper desk. It turned out to be an ideal spot in which to do some work, for it rendered the idea of writing so unlikely as to make it possible again. Objectively good places to work rarely end up being so; in their faultlessness, quiet and well-equipped studies have a habit of rendering the fear of failure overwhelming. Original thoughts are like shy animals. We sometimes have to look the other way – towards a busy street or terminal – before they run out of their burrows.

The setting was certainly rich in distractions. Every few minutes, a voice (usually belonging to either Margaret or her colleague Juliet, speaking from a small room on the floor below) would make an announcement attempting, for example, to reunite a Mrs Barker, recently arrived from Frankfurt, with a stray piece of her hand luggage or reminding Mr Bashir of the pressing need for him to board his flight to Nairobi.

As far as most passengers were concerned, I was an airline employee and therefore a potentially useful source of information on where to find the customs desk or the cash machine. However, those who took the trouble to look at my name badge soon came to regard my desk as a confessional.

One man came to tell me that he was embarking on what he wryly termed the holiday of a lifetime to Bali with his wife, who was just months away from succumbing to incurable brain cancer. She rested nearby, in a specially constructed wheelchair laden with complicated breathing apparatus. She was forty-nine years old and had been entirely healthy until the previous April, when she had gone to work on a Monday morning complaining of a slight headache. Another man explained that he had been visiting his wife and children in London, but that he had a second family in Los Angeles who knew nothing about the first. He had five children in all, and two mothers-in-law, yet his face bore none of the strains of his situation.

Each new day brought such a density of stories that my sense of time was stretched. It seemed like weeks, though it was in fact just a couple of days, since I had met Ana D'Almeida and Sidonio Silva, both from Angola. Ana was headed for Houston, where she was studying business, and Sidonio for Aberdeen, where he was completing a PhD in mechanical engineering. We spent an hour together, during which they spoke in idealistic and melancholy ways of the state of their country. Two days

later, Heathrow held no memories of them, but I felt their absence still.

There were some more permanent fixtures in the terminal. My closest associate was Ana-Marie, who cleaned the section of the check-in area where my desk had been set up. She said she was eager to be included in my book and stopped by several times to chat with me about the possibility. But when I assured her that I would write something about her, a troubled look came over her face and she insisted that I would have to disguise her real name and features. The truth would disappoint too many of her friends and relatives back in Transylvania, she said, for as a young woman she had been the leading student in her conservatoire and since then was widely thought to have achieved renown abroad as a classical singer.

The presence of a writer occasionally raised expectations that something dramatic might be on the verge of occurring, the sort of thing one could read about in a novel. My explanation that I was merely looking around, and required nothing more extraordinary of the airport than that it continue to operate much as it did every other day of the year, was sometimes greeted with disappointment. But the writer's desk was at heart an open invitation to users of the terminal to begin studying

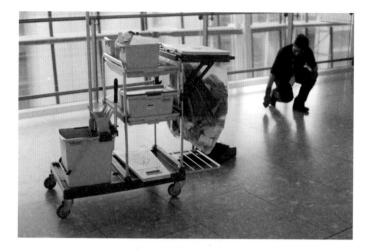

their setting with a bit more imagination and attention, to give weight to the feelings that airports provoke, but which we are seldom able to sort through or elaborate upon in the anxiety of making our way to the gate.

My notebooks grew thick with anecdotes of loss, desire and expectation, snapshots of travellers' souls on their way to the skies – though it was hard to dismiss a worry about what a modest and static thing a book would always be next to the chaotic, living entity that was a terminal.

11   At moments when I could not make headway with my writing, I would go and chat to Dudley Masters, who was based on the floor below me and had spent thirty years cleaning shoes at the airport. His day began at 8.30 a.m. and, around sixty pairs later, finished at 9.00 p.m.

I admired the optimism with which Dudley confronted every new pair of shoes that paused at his station. Whatever their condition, he imagined the best for them, remedying their abuses with an armoury of brushes, waxes, creams and spray cleaners. He knew it was not evil that led people to go for eight months without applying even an all-purpose clear cream polish. He was like a kindly dentist who, on bringing down the ceiling-mounted halogen lamp and asking new patients to open their mouths ('Let's have a look in here, shall we?'), remains aware of how complicated lives can become and so how easily people may give up flossing their teeth while they try to save their companies or minister to a dying parent.

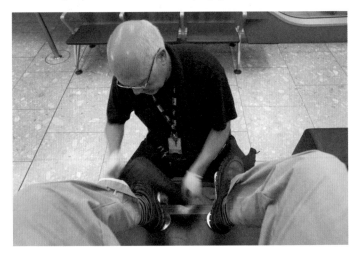

Though he was being paid to shine shoes, he knew that his real mission was psychological. He understood that people rarely have their shoes cleaned at random: they do so when they want to draw a line under the past, when they hope that an outer transformation may be a spur to an inner one. With no ill will, nor any desire to taunt me, he would daily assure me that if he ever got around to putting his experiences down on paper, his would be the most fascinating book about an airport that anyone had ever read.

12 Just past Dudley's workstation, off a corridor leading to the security zone, there was a multi-faith room, a cream-coloured space holding an ill-matched assortment of furniture and a bookshelf of the sacred texts.

I watched a family from southern India coming to pay their respects to Ganesh, the Hindu god in charge of the fortunes of travellers, before going on to board the 1.00 p.m. BA035 flight to Chennai. The deity was presented with some cupcakes and a rose-scented candle, which airport regulations prevented the family from actually lighting.

In the old days, when aircraft routinely fell out of the sky because large and obvious components failed – the fuel pumps gave out or the engines exploded – it felt sensible to cast aside the claims of organised religions in favour of a trust in science. Rather than praying, the urgent task was to study the root causes of malfunctions and stamp out error through reason. But as aviation has become ever more subject to scrutiny, as every part has been hedged by backup systems, so, too, have the reasons for becoming superstitious paradoxically increased.

The sheer remoteness of a catastrophic event occurring invites us to forgo scientific assurances in favour of a more humble stance towards the dangers which our feeble minds struggle to contain. While never going so far as to ignore maintenance schedules, we may nevertheless judge it far from unreasonable to take a few moments before a journey to fall to our knees and pray to the mysterious forces of fate to which all aircraft remain subject and which we might as well call Isis, God, Fortuna or Ganesh – before going on to buy cigarettes and Chanel No.5 in the World Duty Free emporium on the other side of security.

III Airside

1 The security line was impressive as always, numbering at least a hundred people reconciled, though with varying degrees of acceptance, to the idea of not doing very much else with the next twenty minutes of their lives.

The station furthest to the left was staffed by Jim at the scanner, Nina at the manual bag check and Balanchandra at the metal detector. Each had submitted to an arduous year-long course, the essential purpose of which was to train them to look at every human being as though he or she might want to blow up an aircraft – a thoroughgoing reversal of our more customary impulse to find common ground with new acquaintances. The team had been taught to overcome all prejudices as to what an enemy might look like: it could well be the six-year-old girl holding a carton of apple juice and her mother's hand or the frail grandmother flying to Zurich for a funeral. Suspects, guilty until proved innocent, would therefore need to be told in no uncertain terms to step aside from their belongings and stand straight up against the wall.

Like thriller writers, the security staff were paid to imagine life as a little more eventful than it customarily manages to be. I felt sympathy for them in their need to remain alert at every moment of their careers, perpetually poised to react to the most remote of possibilities, of the sort that occurred globally in their line of work perhaps only once in a decade, and even then probably in Larnaca or Baku. They were like members of an evangelical sect living in a country devoid of biblical precedents – Belgium, say, or New Zealand – whose beliefs had inspired a daily expectation of a local return of the Messiah, a prospect not to be discounted even at 3.00 p.m. on a Wednesday in suburban Liège. How enviously the staff must have considered the lot of ordinary policemen and women, who, despite their often unsociable hours and wearying foot patrols, could at least look forward to having regular encounters with exactly the sorts of characters whom they had been trained to deal with.

I felt additional sympathy for the staff as a result of the limited curiosity they were permitted to bring to bear on the targets of their searches. Despite having free rein to look inside any passenger's make-up bag, diary or photo album, they were allowed to investigate only evidence pointing to the presence of explosive devices or murder weapons. There was therefore no sanction for them to ask for whom a neatly wrapped package of underwear was intended, nor any official recognition of how tempting it might occasionally seem to stroke the back pockets of a pair of low-slung jeans without any desire to discover a semi-automatic pistol.

So great was the pressure imposed on the team by the need for vigilance that they were granted more frequent tea breaks than other employees. Every hour they would repair to a room fitted out with dispensing machines, frayed armchairs and pictures of the world's most-wanted terrorists, a series of angry-looking, prophet-like figures with long beards and inscrutable eyes, apparently holed up in mountain caves and reluctant ever to venture into Terminal 5.

It was in this room that I spotted two women who looked as if they might be students enrolled in some sort of internship programme. When I smiled at them, hoping thereby to make them feel a bit more welcome, they came over to greet me and introduced themselves as the two most senior security officers in the building. In charge of training for the entire security staff at Terminal 5, Rachel and Simone regularly taught teams how to disarm terrorists and what positions to adopt in order to protect themselves in the event of a grenade being thrown. They also gave individual employees basic instruction in the use of semi-automatic weapons. Their close focus on anti-terrorism seemed to colour all aspects of their lives: in their spare time, they both read whatever literature they could find on the subject. Rachel was a specialist in the 1976 Entebbe operation, Simone a keen student of the Hindawi Affair, in which a Jordanian man, Nezar Hindawi, had given a Semtex-filled bag to his pregnant girlfriend and persuaded her to board an El Al plane for Tel Aviv. Though the plot had failed, Simone explained (unknowingly damning my naïve conclusions on the wisdom of bothering to search certain sorts of passengers), the incident had forever changed the way security personnel the world over would look at pregnant women, small children and kindly grandmothers.

If many passengers became anxious or angry upon being questioned or searched, it was because such investigations could easily begin to feel, if only on a subconscious level, like accusations, and might thereby slot into pre-existing proclivities towards a sense of guilt.

A long wait for a scanning machine can induce many of us to start asking ourselves if we have perhaps after all left home with an explosive device hidden in our case, or unwittingly submitted to a months-long terrorist training course. The psychoanalyst

Melanie Klein, in her *Envy and Gratitude* (1963), traced this latent sense of guilt back to an intrinsic part of human nature, originating in our Oedipal desire to murder our same-sex parent. So strong can the guilty feeling become in adulthood that it may provoke a compulsion to make false confessions to those in authority, or even to commit actual crimes as a means of gaining a measure of relief from an otherwise overwhelming impression of having done something wrong.

Safe passage through security did have one advantage, at least for those plagued (like the author) by a vague sense of their own culpability. A noiseless, unchecked progress through the detectors allowed one to advance into the rest of the terminal with a feeling akin to that one may experience on leaving church after confession or synagogue on the Day of Atonement, momentarily absolved and relieved of some of the burden of one's sins.

2   There was a good deal of shopping to be done on the other side
    of security, where more than one hundred separate retail outlets
    vied for the attention of travellers – a considerably greater
    number than were to be found in the average shopping centre.
    This statistic regularly caused critics to complain that Terminal
    5 was more like a mall than an airport, though it was hard to
    determine what might be so wrong with this balance, what
    precise aspect of the building's essential aeronautical identity
    had been violated or even what specific pleasure passengers had
    been robbed of, given that we are inclined to visit malls even
    when they don't provide us with the additional pleasure of a
    gate to Johannesburg.

    At the entrance to the main shopping zone was a currency-
exchange desk. Although we are routinely informed that we live
in a vast and diverse world, we may do little more than nod
distractedly at this idea until the moment comes when we find
ourselves at the back of a bureau de change lined with a hundred
safe-deposit boxes, some containing neat sheaves of Uruguayan
pesos, Turkmenistani manats and Malawian kwachas. The trading
desks of the City of London might perform their transactions
with incomparable electronic speed, but patient physical

contact with thick bundles of notes offered a very different sort of immediacy: a living sense of the miscellany of the human species. These notes, in every colour and font, were decorated with images of strongmen, dictators, founding fathers, banana trees and leprechauns. Many were worn and creased from heavy use. They had helped to pay for camels in Yemen or saddles in Peru, been stashed in the wallets of elderly barbers in Nepal or under the pillows of schoolboys in Moldova. A fraying fifty-kina note from Papua New Guinea (bird of paradise on the back, Prime Minister Michael Somare on the front) hardly hinted at the sequence of transactions (from fruit to shoes, guns to toys) that had culminated in its arrival at Heathrow.

Across the way from the exchange desk was the terminal's largest bookshop. Seemingly in spite of the author's defensive predictions about the commercial future of books (perhaps linked to the unavailability of any of his titles at any airport outlet), sales here were soaring. One could buy two volumes and get a third for free, or pick up four and be eligible for a fizzy drink. The death of literature had been exaggerated. Whereas on dating websites, those who like books are usually bracketed into a single category, the broad selections on offer at WH Smith spoke to the diversity of individuals' motives for reading. If there was a conclusion to be drawn from the number of bloodstained covers, however, it was that there was a powerful desire, in a wide cross-section of airline passengers, to be terrified. High above the earth, they were looking to panic about being murdered, and thereby to forget their more mundane fears about the success of a conference in Salzburg or the challenges of having sex for the first time with a new partner in Antigua.

I had a chat with a manager named Manishankar, who had been working at the shop since the terminal first opened. I explained – with the excessive exposition of a man spending a lonely week at the airport – that I was looking for the sort of books in which a genial voice expresses emotions that the reader has long felt but never before really understood; those that convey the secret, everyday things that society at large prefers to leave unsaid; those that make one feel somehow less alone and strange.

Manishankar wondered if I might like a magazine instead. There was no shortage, including several with feature articles on how to look good after forty – advice of course predicated on the assumption that one's appearance had been pleasing at thirty-nine (the writer's age).

Nearby, another bookcase held an assortment of classic novels, which had been imaginatively arranged, not by author or title, but according to the country in which their narratives were set. Milan Kundera was being suggested as a guide to Prague, and Raymond Carver depended upon to reveal the hidden character of the small towns between Los Angeles and Santa Fe. Oscar Wilde once remarked that there had been less fog in London before James Whistler started to paint, and one wondered if the silence and sadness of isolated towns in the American West had not been similarly less apparent before Carver began to write.

Every skilful writer foregrounds notable aspects of experience, details that might otherwise be lost in the mass of data that continuously bathes our senses – and in so doing prompts us to find and savour these in the world around us. Works of literature could be seen, in this context, as immensely subtle instruments by means of which travellers setting out from Heathrow might be urged to pay more careful attention to such things as the conformity and corruption of Cologne society (Heinrich Böll), the quiet eroticism of provincial Italy (Italo Svevo) or the melancholy of Tokyo's subways (Kenzaburō Ōe).

3   It was only after several days of frequenting the shops that I started to understand what those who objected to the dominance of consumerism at the airport might have been complaining about. The issue seemed to centre on an incongruity between shopping and flying, connected in some sense to the desire to maintain dignity in the face of death.

    Despite the many achievements of aeronautical engineers over the last few decades, the period before boarding an aircraft is still statistically more likely to be the prelude to a catastrophe than a quiet day in front of the television at home. It therefore tends to raise questions about how we might best spend the last moments before our disintegration, in what frame of mind we might wish to fall back down to earth – and the extent to which we would like to meet eternity surrounded by an array of duty-free bags.

    Those who attacked the presence of the shops might in essence have been nudging us to prepare ourselves for the end.

At the Blink beauty bar, I felt anew the relevance of the traditional religious call to seriousness voiced in Bach's Cantata 106:

> *Bestelle dein Haus,*
> *Denn du wirst Sterben,*
> *Und nicht lebendig bleiben.*
>
> Set thy house in order,
> For thou shalt die,
> And not remain alive.

Despite its seeming mundanity, the ritual of flying remains indelibly linked, even in secular times, to the momentous themes of existence – and their refractions in the stories of the world's religions. We have heard about too many ascensions, too many voices from heaven, too many airborne angels and saints to ever be able to regard the business of flight from an entirely pedestrian perspective, as we might, say, the act of travelling by train. Notions of the divine, the eternal and the significant accompany us covertly on to our craft, haunting the reading aloud of the safety instructions, the weather announcements made by our captains and, most particularly, our lofty views of the gentle curvature of the earth.

4 It seemed appropriate that I should bump into two clergymen just outside a perfume outlet, which released the gentle, commingled smell of some eight thousand varieties of scent. The older of the pair, the Reverend Sturdy, wore a high-visibility jacket with the words 'Airport Priest' printed on the back. In his late sixties, he had a vast and archetypically ecclesiastical beard and gold-rimmed spectacles. The cadence of his speech was impressively slow and deliberate, like that of a scholar unable to ignore, even for a moment, the nuances behind every statement, and accustomed to living in environments where these could be investigated to their furthest conclusions without fear of inconveniencing or delaying others. His colleague, Albert Kahn, likewise garbed for high visibility – though his jacket, borrowed from another staff member, read merely 'Emergency Services' – was in his early twenties and on a work placement at Heathrow while completing theological studies at Durham University.

'What do people tend to come to you to ask?' I enquired of the Reverend Sturdy as we passed by an outlet belonging to that perplexingly indefinable clothing brand Reiss. There was a long pause, during which a disembodied voice reminded us once more never to leave our luggage unattended.

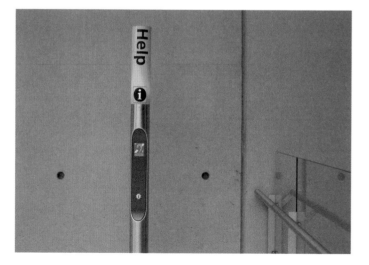

'They come to me when they are lost,' the Reverend replied at last, emphasising the final word so that it seemed to reflect the spiritual confusion of mankind, a hapless race of beings described by St Augustine as 'pilgrims in the City of Earth until they can join the City of God'.

'Yes, but what might they be feeling lost *about?*'

'Oh,' said the Reverend with a sigh, 'they are almost always looking for the toilets.'

Because it seemed a pity to end our discussion of metaphysical matters on such a note, I asked the two men to tell me how a traveller might most productively spend his or her last minutes before boarding and take-off. The Reverend was adamant: the task, he said, was to turn one's thoughts intently to God.

'But what if one can't believe in him?' I pursued.

The Reverend fell silent and looked away, as though this were not a polite question to ask of a priest. Happily, his colleague, weaned on a more liberal theology, delivered an equally succinct but more inclusive reply, to which my thoughts often returned in the days to come as I watched planes taxiing out to the runways: 'The thought of death should usher us towards whatever happens to matter most to us; it should lend us the courage to pursue the way of life we value in our hearts.'

Just beyond the security area was a suite, named after an ill-fated supersonic jet and reserved for the use of first-class passengers. The advantages of wealth can sometimes be hard to see: expensive cars and wines, clothes and meals are nowadays rarely proportionately superior to their cheaper counterparts, due to the sophistication of modern processes of design and mass production. But in this sense, British Airways' Concorde Room was an anomaly. It was humblingly and thought-provokingly nicer than anywhere else I had ever seen at an airport, and perhaps in my life.

There were leather armchairs, fireplaces, marble bathrooms, a spa, a restaurant, a concierge, a manicurist and a hairdresser. One waiter toured the lounge with plates of complimentary caviar, foie gras and smoked salmon, while a second made circuits with éclairs and strawberry tartlets.

'For what purpose is all the toil and bustle of this world? What is the end of the pursuit of wealth, power and pre-eminence?' asked Adam Smith in *The Theory of Moral Sentiments* (1759), going on to answer, 'To be observed, to be attended to, to be taken notice of with sympathy, complacency, and approbation' – a set of ambitions to which the creators of the Concorde Room had responded with stirring precision.

As I took a seat in the restaurant, I felt certain that whatever it had taken for humanity to arrive at this point had ultimately been worth it. The development of the combustion engine, the invention of the telephone, the Second World War, the introduction of real-time financial information on Reuters screens, the Bay of Pigs, the extinction of the slender-billed curlew – all of these things had, each in its own fashion, helped to pave the way for a disparate group of uniformly attractive individuals to silently mingle in a splendid room with a view of a runway in a cloud-bedecked corner of the Western world.

'There is no document of culture that is not at the same time a document of barbarism,' the literary critic Walter Benjamin had once famously written, but that sentiment no longer seemed to matter very much.

Still, I recognised the fragility of the achievement behind the lounge. I sensed how relatively few such halcyon days there might be left before members of the small fraternity ensconced in its armchairs came to grief and its gilded ceilings cracked into ruin. Perhaps it had felt a bit like this on the terraces of Hadrian's villa outside Rome on autumn Sunday evenings in the second century AD, as a blood-red sun set over the marble colonnades. One might have had a similar presentiment of catastrophe, looming in the form of the restless Germanic tribes lying in wait deep in the sombre pine forests of the Rhine Valley.

I started to feel sad about the fact that I might not be returning to the Concorde Room anytime soon. I realised, however, that the best way to attenuate my grief would be to nurture a thoroughgoing hatred of all those more regularly admitted into the premises. Over a plate of porcini mushrooms on a brioche base, I therefore tried out the idea that the lounge was really a hideout for a network of oligarchs who had won undeserved access through varieties of nepotism and skulduggery.

Regrettably, on closer examination, I was forced to concede that the evidence conflicted unhelpfully with this otherwise consoling thesis, for my fellow guests fitted none of the stereotypes of the rich. Indeed, they stood out chiefly on the basis of how *ordinary* they looked. These were not the chinless heirs to hectares of countryside but rather normal people who had figured out how to make the microchip and spreadsheet work on their behalf. Casually dressed, reading books by Malcolm Gladwell, they were an elite who had come into their wealth by dint of intelligence and stamina. They worked at Accenture fixing irregularities in supply chains or built income-ratio models at MIT; they had started telecommunications companies or did astrophysical research at the Salk Institute. Our society is affluent in large part because its wealthiest citizens do not behave the way rich people are popularly supposed to. Simple plunder could never have built up this sort of lounge (globalised, diverse, rigorous, technologically-minded), but at best a few gilded pleasure palaces standing out in an otherwise feudal and backward landscape.

In the rarefied air that was pumped into the Concorde Room, there nonetheless hovered a hint of something troubling: the implicit suggestion that the three traditional airline classes represented nothing less than a tripartite division of society according to people's genuine talents and virtues. Having abolished the caste systems of old and fought to ensure universal access to education and opportunity, it seemed that we might have built up a meritocracy that had introduced an element of true justice into the distribution of wealth as well as of poverty. In the modern era, destitution could therefore be regarded as not merely pitiable but *deserved*. The question of why, if one was in any way talented or adept, one was still unable to earn admittance to an elegant lounge was a conundrum for all economy airline passengers to ponder in the privacy of their own minds as they perched on hard plastic chairs in the overcrowded and chaotic public waiting areas of the world's airports.

The West once had a powerful and forgiving explanation for exclusion from any sort of lounge: for two thousand years Christianity rejected the notion, inherent in the modern meritocratic system, that virtue must inevitably usher in material success. Jesus was the highest man, the most blessed, and yet throughout his earthly life he was poor, thus by his very example ruling out any direct equation between righteousness and wealth. The Christian story emphasised that, however apparently equitable our educational and commercial infrastructures might seem, random factors and accidents would always conspire to wreck any neat alignment between the hierarchies of wealth on the one hand and virtue on the other. According to St Augustine, only God himself knew what each individual was worth, and He would not reveal that assessment before the time of the Last Judgement, to the sound of thunder and the trumpets of angels – a phantasmagorical scenario for non-believers, but helpful nevertheless in reminding us to refrain from judging others on the basis of a casual look at their tax returns.

The Christian story has neither died out nor been forgotten. That it continues even now to scratch away at meritocratic explanations of privilege was made clear to me when, after a copious lunch rounded off by a piece of chocolate cake with passionfruit sorbet, an employee called Reggie described for me the complicated set of circumstances that had brought her to the brutally decorated staff area of the Concorde Room from a shantytown outside Puerto Princesa in the Philippines. Our preference for the meritocratic versus the Christian belief system will in the end determine how we decide to interpret the relative standing of a tracksuited twenty-seven-year-old entrepreneur reading the *Wall Street Journal* by a stone-effect fireplace while waiting to board his flight to Seattle, against that of a Filipina cleaner whose job it is to tour the bathrooms of an airline's first-class lounge, swabbing the shower cubicles of their diverse and ever-changing colonies of international bacteria.

Although the majority of its users regarded it as little more than a place where they had to spend a few hours on their way to somewhere else, for many others the terminal served as a permanent office, one that accommodated a thousand-strong bureaucracy across a series of floors off limits to the general public. The work done here was not well suited to those keen on seeing their own identities swiftly or flatteringly reflected back at them through their labour. The terminal had taken some twenty years and half a million people to build, and now that it was finally in operation, its business continued to proceed ponderously and only by committee. Layer upon layer of job titles (Operational Resource Planning Manager, Security Training and Standards Adviser, Senior HR Business Partner) gave an indication of the scale of the hierarchies that had to be consulted before a new computer screen could be acquired or a bench repositioned.

A few of the more obscure offices nonetheless managed to convey an impressive sense of the scope of the manpower and intelligence involved in getting planes around the world. The area housing British Airways' Customer Experience Division was filled with prototypes of cabin seats, life jackets, vomit bags, mints and towelettes. An archivist oversaw a room filled with rejected samples, most of which had ended up there on the grounds of cost, not so much because of the airline's miserliness as because of its sheer size, an overspend on a single chair having dramatic consequences when typing out a purchase order for thirty thousand of them. A tour of the premises, with a close look at the early designs for plane interiors, offered a pleasure similar to that of looking through the first draft of a manuscript and seeing that prose that would eventually be polished and sure had started out hesitant and confused – a lesson with consoling applications for a universal range of maiden efforts.

I came away from the back rooms of the airport regretting what seemed a wrong-headed imbalance between the lavish attention paid to distracting and entertaining travellers and the scant time spent educating them about the labour involved in their journeys.

It is a good deal more interesting to find out how an airline meal is made than how it tastes – and a good deal more troubling. A mile from the terminal, in a windowless refrigerated factory owned by the Swiss company Gate Gourmet, eighty thousand breakfasts, lunches and dinners, all intended for ingestion within the following fifteen hours somewhere in the troposphere, were being made up by a group of women from Bangladesh and the Baltic. Korean Airlines would be serving beef broth, JAL salmon teriyaki and Air France a chicken escalope on a bed of puréed carrots. Foods that would later be segregated according to airline and destination now mingled freely together, like passengers in the terminal, so that a tray containing a thousand plates of Dubai-bound Emirates hummus might be lined up in the freezer room next to four trolleys full of SAS gravadlax, set to fly part-way to Stockholm.

Aeroplane food stands at a point of maximum tension between the man-made and the natural, the technological and the organic. Even the most anaemic tomato (and the ones at Gate Gourmet were mesmerising in their fibrous pallor) remains a work of nature. How strange and terrifying, then, that we should take our fruit and vegetables up into the sky with us, when we used to sit more humbly at nature's feet, hosting harvest festivals to honour the year's wheat crop and sacrificing animals to ensure the continued fecundity of the earth.

There is no need for such prostration now. A batch of twenty thousand cutlets, which had once, if only briefly, been attached to lambs born and nursed on Welsh hillsides, was driven into the depot. Within hours, with the addition of a breadcrumb

topping, a portion of these would metamorphose into meals that would be eaten over Nigeria – with no thought or thanks given to their author, twenty-six-year-old Ruta from Lithuania.

7 The British Airways flight crews also maintained offices at the airport. In an operations room in Terminal 5, pilots stopped by throughout the day and into the evening to consult with their managers about what the weather was like over Mongolia, or how much fuel they ought to purchase in Rio. When I saw an opening, I introduced myself to Senior First Officer Mike Norcock, who had been flying for fifteen years and who greeted me with one of those wry, indulgent smiles often bestowed by professionals upon people with a more artistic calling. In his presence, I felt like a child unsure of his father's affections. I realised that meeting pilots was doomed to escalate into an ever more humiliating experience for me, as the older I got, the more obvious it became that I would never be able to acquire the virtues that I so admired in them – their steadfastness, courage, decisiveness, logic and relevance – and must instead forever remain a hesitant and inadequate creature who would almost certainly start weeping if asked to land a 777 amid foggy ground conditions in Newfoundland.

Norcock had come to the operations room to pick up some route maps. He was off to India in his jumbo but first wanted to double-check the weather over Iran's northern border. He knew so much that his passengers did not. He understood, for example, that the sky, which we laypeople so casually and naïvely tend to appraise in terms of its colour and cloud formations, was in fact criss-crossed by coded flight lanes, intersections, junctions and beacon signals. On this day, he was especially concerned with VAN115.2, both a small orange dot on the flight charts and a wooden shed two metres high and five across, situated on the edge of a farmer's field at the top of a gorge in a thinly inhabited part of eastern Turkey – a location where Norcock would in a few hours be taking a left fork on to airway R659, as his passengers anxiously anticipated their lunch, a lasagne being prepared even now in Gate Gourmet's factory. I looked at his steady, well-sculpted hands and thought of how far he had come since childhood.

I knew, at least in theory, that Norcock could not always, in every circumstance, be a model of authoritative and patriarchal behaviour. He, too, must be capable of petulance, of vanity, of acting foolishly, of making casually cruel remarks to his spouse or neglecting to understand his children. There are no directional charts for daily life. But at the same time, I was reluctant to either accept or exploit the implications of this knowledge. I wanted to believe in the capacity of certain professions to enable us to escape the ordinary run of our frailties and to accede, if only for a moment, to a more impressive sort of existence than most of us will ever know.

From the outset, my employer had suggested that I might wish to conduct a brief interview with one of the most powerful men in the terminal: the head of British Airways, Willie Walsh. It was a daunting prospect, as Walsh was having a busy time of it. His company was losing an average of £1.6 million a day, a total of £148 million over the previous three months. His pilots and cabin crew were planning strikes. Studies showed that his baggage handlers misappropriated more luggage than their counterparts at any other European airline. The government wanted to tax his fuel. Environmental activists had been chaining themselves to his fences. He had infuriated those in the upper echelons at Boeing by telling them that he would not be able to keep up with the prepayment schedule he had committed to for the new 787 aircraft he had ordered. His efforts to merge his airline with Qantas and Iberia had stalled. He had done away with the free chocolates handed round after every meal in business class, and in the process provoked a three-day furore in the British press.

Journalism has long been enamoured of the idea of the interview, beneath which lies a fantasy about access: a remote figure, beyond the reach of the ordinary public and otherwise occupied with running the world, opens up and reveals his innermost self to a correspondent. With admission set at the price of a newspaper, the audience is invited to forget their station in life and accompany the interviewer into the palace or the executive suite. The guards lay down their weapons, the secretaries wave the visitors through. Now we are in the inner sanctum. While waiting, we have a look around. We learn that the president likes to keep a bowl of peppermints on his desk, or that the leading actress has been reading Dickens.

But the tantalising promise of shared secrets is rarely fulfilled as we might wish, for it is almost never in the interests of a prominent figure to become intimate with a member of the press. He has better people on to whom to unburden himself. He does not need a new friend. He is not going to disclose his plots for vengeance or his fears about his professional future. For the

celebrity, the interview is thus generally reduced to an exercise in saying as little as possible without confounding the self-love of the journalist on the sofa, who might become dangerous if rendered too starkly aware of the futility of his mission. In a bid to appease the underlying demand for closeness, the subject may let it drop that he is about to go on holiday to Florida, or that his daughter is learning how to play tennis.

There was evidently nothing of standard consequence that I could ask Mr Walsh. There was no point in my bringing up pensions, carbon emissions, premium yields or even the much-missed chocolates – no point, really, in our meeting at all, had not events reached the stage where articulating this insight would have seemed rude.

So we got together for forty minutes in a conference room, between Mr Walsh's meeting with a trade-union representative and a delegation from Airbus. I felt as if I were interrupting a discussion of beachheads between Roosevelt and Churchill in May 1943.

Fortunately, I had come to the conclusion that though Mr Walsh was the CEO of one of the world's largest airlines, it would be wholly unfair of me to treat him like a businessman.

The fiscal state of his company was simply too precarious, and too woefully inaccurate a reflection of his talents and interests, to permit me to confuse Mr Walsh with his balance sheet.

Considered collectively, as a cohesive industry, civil aviation had never in its history shown a profit. Just as significantly, neither had book publishing. In this sense, then, the CEO and I, despite our apparent differences, were in much the same sort of business, each one needing to justify itself in the eyes of humanity not so much by its bottom line as by its ability to stir the soul. It seemed as unfair to evaluate an airline according to its profit-and-loss statement as to judge a poet by her royalty statements. The stock market could never put an accurate price on the thousands of moments of beauty and interest that occurred around the world every day under an airline's banner: it could not describe the sight of Nova Scotia from the air, it had no room in its optics for the camaraderie enjoyed by employees in the Hong Kong ticket office, it had no means of quantifying the adrenalin-thrill of take-off.

The logic of my argument was not lost on Mr Walsh, who had himself once been a pilot. As we talked, he expressed his admiration for the way planes, vast and complicated machines, could defy their size and the challenges of the atmosphere to soar into the sky. We remarked on the surprise we both felt on seeing a 747 at a gate, dwarfing luggage carts and mechanics, at the idea that such a leviathan could move – a few metres' distance, let alone across the Himalayas. We reflected on the pleasure of seeing a 777 take off for New York and, over the Staines reservoir, retract its flaps and wheels, which it would not require again until its descent over the white clapboard houses of Long Beach, some 5,000 kilometres and six hours of sea and cloud away. We exclaimed over the beauty of a crowded airfield, where, through the heat haze of turbofans, the interested observer can make out sequences of planes waiting to begin their journeys, their fins a confusion of colours against the grey horizon, like sails at a regatta. In another life, I decided, the chief executive and I might have become good friends.

We were getting on so well that Mr Walsh – or Willie, as he now urged me to call him – suggested we repair to the lobby downstairs, where we could have a look at a model of the new A380, twelve of which he had ordered from Airbus and which would be joining the British Airways fleet in 2012. Once we were standing before it, Willie, with what seemed a child's sense of delight, invited me to join him in climbing up on to a bench to appreciate the sheer scale of the jet's ailerons and the breadth of its fuselage.

So much warmth did I feel for him as we stood shoulder to shoulder, admiring his model plane, that I was emboldened to mention a fantasy I had harboured since I first received authorisation to write a book about Heathrow. I asked Willie whether, if he had any money left, he might one day consider appointing me his writer-in-flight, in order that I might constantly circumnavigate the earth composing, among other things, sincere dedications to my patron, impressionistic essays describing the ochre colours of the western Australian desert as seen from the flight deck, and vignettes recounting the balletic routines of the stewards in the galley.

There was a pause, and for a moment the bonhomie disappeared from the chief executive's handsome grey-green eyes. But soon enough it returned. 'Of course,' he said, beaming. 'Once at Aer Lingus, the video system broke down, and we invited a couple of Irish minstrels to sing songs on a flight to New York. Alan, I could see you at the front of the cabin doing a ditty or two for our passengers.' And following that prognostication, he apologised for taking up so much of my precious time and called for a security officer to escort me to the door of his corporate headquarters.

Not long into my stay, evening became my favourite time at the airport. By eight, most of the choppy short-haul European traffic had come and gone. The terminal was emptying out, Caviar House was selling the last of its sturgeon eggs and the cleaning teams were embarking on the day's most systematic mopping of the floors. Because it was summer, the sun would not set for another forty minutes, and in the interim a gentle, nostalgic light would flood across the seating areas.

The majority of the passengers left in the terminal at this hour were booked on one or another of the flights that departed every evening for the East, unbeknownst to most of the households of north-west London which they crossed en route for Singapore, Seoul, Hong Kong, Shanghai, Tokyo and Bangkok.

The atmosphere in the waiting areas was lonely, but curiously, the feeling was benign for being so general, eliminating the unease that any one individual might otherwise experience at being the only one to be alone, and thus paradoxically making new connections seem possible in a way they might not have done in the more obviously convivial surroundings of a crowded city bar. At night, the airport emerged as a home of nomadic spirits, types who could not commit to any one country, who shied from tradition and were suspicious of settled community, and who were therefore nowhere more comfortable than in the intermediate zones of the modern world, landscapes gashed by kerosene storage tanks, business parks and airport hotels.

Because the arrival of night typically pulls us back towards the hearth, there seemed something especially brave about travellers who were preparing to entrust themselves to the darkness, to be carried in a craft navigated by instruments alone and to surrender to sleep, finally, only over Azerbaijan or the Kalahari Desert.

In a control room beside the terminal, a giant map of the world showed the real-time position of every plane in the British Airways fleet, as tracked by a string of satellites. Across the globe, 180 aircraft were on their way, together holding some one hundred thousand passengers. A dozen planes were crossing the North Atlantic, five were routing around a hurricane to the west of Bermuda, and one could be seen plotting a course over Papua New Guinea. The map was emblematic of a touching vigilance, for however far removed each craft was from its home

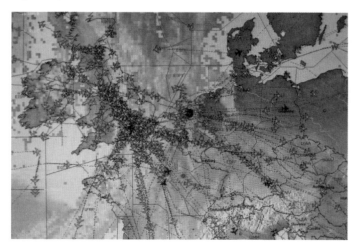

airfield, however untethered and able it looked, it was never far from the minds of those in the control room in London, who, like parents worrying about their children, would not feel at ease until each of their charges had safely touched down.

Every night a few planes would be towed away from their gates to a set of giant hangars, where a phalanx of gangways and cranes would lock themselves around their organically shaped bodies like a series of handcuffs. While aircraft tended to be coy about their need to pay such visits – hardly letting on, at the close of a trip from Los Angeles or Hong Kong, that they had reached the very end of their permitted quota of nine hundred flying hours – the checks provided an opportunity for them to reveal their individuality. What to passengers might have looked like yet another indistinguishable 747 would emerge, during this process, as a machine with a distinct name and medical history: G-BNLH, for example, had come into service in 1990 and in the intervening years had had three hydraulic leaks over the Atlantic, once blown a tyre in San Francisco and, only the previous week, dropped an apparently unimportant part of its wing in Cape Town. Now it was coming into the hangar with, among other ailments, twelve malfunctioning seats, a large smear of purple nail polish on a wall panel and an opinionated

microwave oven in a rear galley that ignited itself whenever an adjacent basin was used.

Thirty men would work on the plane through the night, the whole operation guided by an awareness that, while the craft could under most circumstances be extraordinarily forgiving, a chain of events originating in the failure of something as small as a single valve could nevertheless bring it down, just as a career might be ruined by one incautious remark, or a person die because of a clot less than a millimetre across.

I toured the exterior of the aircraft on a gangway that ran around its midriff and let my hands linger on its nose cone, which had only a few hours earlier carved a path through dense layers of static cumulus clouds.

Studying the plane's tapered tail, and the marks left across the back of its fuselage by the enraged thrust of its four RB211 engines, I wondered if scientists and engineers might have designed planes and their means of take-off differently had our species been graced with some subtler, less thunderous mode of conception, perhaps one managed frictionlessly and quietly by the male's sitting for a few hours on an egg left behind in a leafy recess by the female.

10 At around eleven-fifteen each night, by government decree, the airport was closed to both incoming and outgoing traffic. Across the aprons, all was suddenly as quiet as it must have been a hundred years ago, when there was nothing here but sheep meadows and apple farms. I met up with a man called Terry, whose job it was to tour the runways in the early hours looking out for stray bits of metal. We drove out to a spot at the end of the southern runway, 27L to pilots, which Terry termed the most expensive piece of real estate in Europe. It was here, at forty-second intervals throughout the day, on a patch of tarmac only a few metres square and black with rubber left by tyres, that the aircraft of the world made their first contact with the British Isles. This was the exact set of coordinates that planes anticipated from across southern England: even in the thickest fog, their automatic landing systems could pick up the glide-path beam that was projected up into the sky from this point, the radio wave calling them to place their wheels squarely in the centre of a zone highlighted by a double line of parallel white lights.

But just now, the patch of runway that was almost solely responsible for destroying the peace and quiet of some ten million people was becalmed. One could walk unhurriedly across it and even give in to the temptation to sit cross-legged on its centreline, a gesture that partook of some of the sublime thrill of touching a disconnected high-voltage electricity cable, running one's fingers along the teeth of an anaesthetised shark or having a wash in a fallen dictator's marble bathroom.

A field mouse scurried out of the grass and on to the runway, where for a moment it stood still, transfixed by the jeep's headlamps. It was of a kind which regularly populates children's books, where mice are always clever and good-natured creatures who live in small houses with red-and-white-checked curtains, in sharp contrast to the boorish humans, who are clumsily oversized and unaware of their own limits. Its presence this night on the moonlit tarmac served optimistically to suggest that when mankind is finished with flying – or more generally, with *being* – the earth will retain a capacity to absorb our follies and make way for more modest forms of life.

11 Terry dropped me off at my hotel. I felt too stimulated to sleep –
and so went for a drink at an all-night bar frequented by delayed
flight crews and passengers.

Over a dramatically sized tequila-based cocktail named an
After Burner, I befriended a young woman who told me that
she was writing a doctoral dissertation at the University of
Warsaw: her subject was the Polish poet and novelist Zygmunt
Krasiński, with a particular focus on his famous work *Agaj-Han*
(1834) and the tragic themes explored therein. She argued that
Krasiński's reputation had been unfairly eclipsed during the
twentieth century by that of his fellow Romantic writer Adam
Mickiewicz, and explained that she had been motivated in her
research by a desire to reacquaint her compatriots with an aspect
of their heritage that had been deliberately denied them during
the Communist era. When I asked why she was at the airport,
she replied that she had come to meet a friend from Dubai,
whose plane had been delayed and was now unlikely to land
at Heathrow before mid-morning. An engineer of Lebanese
origin, he had been coming to London once a month for the
past year and a half in order to receive treatment for throat
cancer at a private hospital in Marylebone, and during every
visit, he invited her to spend the night with him in a Prestige

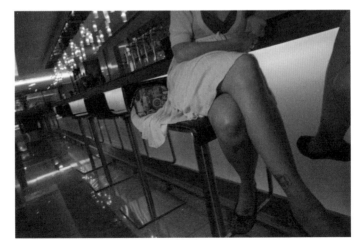

Suite on the top floor of the Sofitel. She confided that she was registered with an agency which had a head office in Hayes and added, in a not unrelated aside, that Zygmunt Krasiński had conducted a three-year-long affair with the Countess Delfina Potocka, with whom Chopin had also been in love.

I returned to my room at three in the morning, struck by a sense of our race as a peculiar, combustible mixture of the beast and the angel. The first plane, due at Heathrow at dawn, was now somewhere over western Russia.

IV   Arrivals

1 There used to be time to arrive. Incremental geographical changes would ease the inner transitions: desert would gradually give way to shrub, savannah to grassland. At the harbour, the camels would be unloaded, a room would be found overlooking the customs house, passage would be negotiated on a steamer. Flying fish would skim past the ship's hull. The crew would play cards. The air would cool.

Now a traveller may be in Abuja on Tuesday and at the end of a satellite in the new terminal at Heathrow on Wednesday. Yesterday lunchtime, one had fried plantain in the Wuse District to the sound of an African cuckoo, whereas at eight this morning the captain is closing down the 777's twin engines at a gate next to a branch of Costa Coffee.

Despite one's exhaustion, one's senses are fully awake, registering everything – the light, the signage, the floor polish, the skin tones, the metallic sounds, the advertisements – as sharply as if one were on drugs, or a newborn baby, or Tolstoy. Home all at once seems the strangest of destinations, its every detail relativised by the other lands one has visited. How peculiar this morning light looks against the memory of dawn in the Obudu hills, how unusual the recorded announcements sound after the wind in the High Atlas and how inexplicably English

(in a way they will never know) the chat of the two female ground staff seems when one has the din of a street market in Lusaka still in one's ears.

One wants never to give up this crystalline perspective. One wants to keep counterpoising home with what one knows of alternative realities, as they exist in Tunis or Hyderabad. One wants never to forget that nothing here is normal, that the streets are different in Wiesbaden and Luoyang, that this is just one of many possible worlds.

2   In the brief history of aviation, not many airports have managed to fulfil their visitors' hopes for an architecture that might properly honour the act of arrival. Too few have followed the example set by Jerusalem's elaborate Jaffa Gate, which once welcomed travellers who had completed the journey to the Holy City across the baked Shephelah plains and through the thief-infested Judean hills. But Terminal 5 wanted to have a go.

In the older terminals at Heathrow, it was a certain sort of carpet that one tended to notice first, swirling green, yellow, brown and orange, around which there hovered associations of vomit, pubs and hospitals. Here, by way of contrast, there were handsome grey composite tiles, bright corridors lined with glass panels in a calming celadon shade and bathrooms fitted out with gracious sanitary ware and full-length cubicle doors made of heavy timber.

The structure was proposing a new idea of Britain, a country that would be reconciled to technology, that would no longer be in thrall to its past, that would be democratic, tolerant, intelligent, playful and lacking in spite or irony. All this was a simplification, of course: twenty kilometres to the west and north were tidy hamlets and run-down estates that would at once have contravened any of the suggestions encoded in the terminal's walls and ceilings.

Nevertheless, like Geoffrey Bawa's Parliament in Colombo or Jørn Utzon's Opera House in Sydney, Richard Rogers's Terminal 5 was applying the prerogative of all ambitious architecture to create rather than merely reflect an identity. It hoped to use the hour or so when passengers were within its space – objectively, to have their passports stamped and to recover their luggage – to define what the United Kingdom might one day become, rather than what it too often is.

3 Upon disembarking, after a short walk, arriving passengers entered a hall that tried hard to downplay the full weight of its judicial role. There were no barriers, guns or reinforced booths, merely an illuminated sign overhead and a thin line of granite running across the floor. Power was sure of itself here, confident enough to be restrained and invisible to those privileged, by an accident of birth, to skirt it. Three times a day, a cleaning team came and swept their brooms across the line that marked the divide between the no man's land of the aircraft on the one side and, on the other, the well-stocked pharmacies, benign mosquitoes, generous library lending policies, sewage plants and pelican crossings available to visitors and residents of Great Britain alike.

With just a single unhappy swipe of the computer, however, all such implicit promises might be prematurely broken. A guard would be called and would lead the unfortunate traveller from the immigration hall to a suite of rooms two storeys below. The children's playroom seemed especially poignant in its fittings: there was a Brio train, most of the Lego City range, a box of Caran d'Ache pens and, for each new child sequestered there, a box of snacks and plastic animals, his or hers to keep.

In the imaginations of certain children in Eritrea or Somalia, England would hence always remain a briefly glimpsed country of Quavers, Jelly Tots and squared cartons of orange juice – a country so rich it could afford to give away small digital alarm clocks, and one whose guards knew how to put wooden train tracks together. Next door, in a barer room in which every word was being captured by a police tape recorder, their parents would experience another side of the nation, as they delineated their unsuccessful applications to an impassive member of the immigration service.

4   Over the course of history, few joyful moments can have unfolded in a baggage-reclaim area, though the one in the terminal was certainly doing its best to keep its users optimistic.

It had high ceilings, flawlessly poured concrete walls and trolleys in abundance. Furthermore, the bags came quite quickly. The company responsible for the conveyor belts, Vanderlande Industries from the Netherlands, had made its reputation in the mail-order and parcel-distribution sectors and was now the world leader in suitcase logistics. Seventeen kilometres' worth of conveyor belts ran under the terminal, where they were capable of processing some twelve thousand pieces of luggage an hour. One hundred and forty computers scanned tags, determined where individual bags were going and checked them for explosives along the way. The machines treated the suitcases with a level of care that few humans would have shown them: when the bags had to wait in transit, robots would carry them gently over to a dormitory and lay them down on yellow mattresses, where – like their owners in the lounges above – they would loll until their flights were ready to receive them. By the time they were lifted off the belt, many suitcases were likely to have had more interesting travels than their owners.

Nevertheless, in the end, there was something irremediably melancholic about the business of being reunited with one's luggage. After hours in the air free of encumbrance, spurred on to formulate hopeful plans for the future by the views of coasts and forests below, passengers were reminded, on standing at the carousel, of all that was material and burdensome in existence. There were some elemental dualities at work in the contrasting realms of the baggage-reclaim hall and the aeroplane – dichotomies of matter and spirit, heaviness and lightness, body and soul – with the negative halves of the equations all linked to the stream of almost identical black Samsonite cases that rolled ceaselessly along the tunnels and belts of Vanderlande's exquisite conveyor apparatus.

Around the carousel, as in a Roman traffic jam, trolleys grimly refused to cede so much as a centimetre to one another. Although each suitcase was a repository of dense and likely fascinating individuality – this one perhaps containing a lime-coloured bikini and an unread copy of *Civilization and Its Discontents*, that one a dressing gown stolen from a Chicago hotel and a packet of Roche antidepressants – this was not the place to start thinking about anyone else.

5 Yet the baggage area was only a prelude to the airport's emotional climax. There is no one, however lonely or isolated, however pessimistic about the human race, however preoccupied with the payroll, who does not in the end expect that someone significant will come to say hello at arrivals.

Even if our loved ones have assured us that they will be busy at work, even if they told us they hated us for going travelling in the first place, even if they left us last June or died twelve and a half years ago, it is impossible not to experience a shiver of a sense that they may have come along anyway, just to surprise us and make us feel special (as someone must have done for us when we were small, if only occasionally, or we would never have had the strength to make it this far).

It is therefore hard to know just what expression we should mould our faces into as we advance towards the reception zone. It might be foolhardy to relinquish the solemn and guarded demeneaour we usually adopt while wandering through the anonymous spaces of the world, but at the same time, it seems only right that we should leave open at least the suggestion of a smile. We may settle on the sort of cheerful but equivocal look commonly worn by people listening out for punchlines to jokes narrated by their bosses.

So what dignity must we possess not to show any hesitation when it becomes clear, in the course of a twelve-second scan of the line, that we are indeed alone on the planet, with nowhere to head to other than a long queue at the ticket machine for the Heathrow Express. What maturity not to mind that only two metres from us, a casually dressed young man perhaps employed in the lifeguard industry has been met with a paroxysm of joy by a sincere and thoughtful-looking young woman with whose mouth he is now involved. And what a commitment to reality it will take for us not to wish that we might, just for a time, be not our own tiresome selves but rather Gavin, flying in from

Los Angeles after a gap year in Fiji and Australia, with whose devoted parents, exhilarated aunt, delighted sister, two girl friends and a helium balloon we might therefore repair to a house on the southern outskirts of Birmingham.

At arrivals, there were forms of welcome of which princes would have been jealous, and which would have rendered inadequate the celebrations laid on at Venice's quaysides for the explorers of the Eastern silk routes. Individuals without official status or distinguishing traits, passengers who had sat unobtrusively for twenty-two hours near the emergency exits, now set aside their bashfulness and revealed themselves as the intended targets of flags, banners, streamers and irregularly formed home-baked chocolate biscuits – while, behind them, the chiefs of large corporations prepared for glacial limousine rides to the marble-and-orchid-bedecked lobbies of their luxury hotels.

The prevalence of divorce in modern society guaranteed an unceasing supply of airport reunions between parents and children. In this context, there was no longer any point in pretending to be sober or stoic: it was time to squeeze a pair of frail and yet plump shoulders very tightly and founder into tears. We may spend the better part of our professional lives projecting strength and toughness, but we are all in the end creatures of appalling fragility and vulnerability. Out of the millions of people we live among, most of whom we habitually ignore and are ignored by in turn, there are always a few who hold hostage our capacity for happiness, whom we could recognise by their smell alone and whom we would rather die than be without. There were men pacing impatiently and blankly who had looked forward to this moment for half a year and could not restrain themselves any further at the sight of a small boy endowed with their own grey-green eyes and their mother's cheeks, emerging from behind the stainless-steel gate, holding the hand of an airport operative.

At such moments, it felt almost as if death itself had been averted – and yet there was also a sense, lending the occasion more poignancy still, that it could not go on being cheated for ever. Perhaps this was a way of practising for mortality. Some day, many years from now, the adult child would say goodbye to his father before going on a routine business trip, and the reprieve would abruptly run out. There would be a telephone call in the middle of the night to a room on the twentieth floor of a Melbourne hotel, bringing the news that the parent had suffered a catastrophic seizure on the other side of the world and that there was nothing more the doctors could do for him – and from that day forward, for the now-grown-up boy, the line in arrivals would always be missing one face in particular.

6 Not all meetings were so emotional. One might have come from Shanghai to join Malcolm and Mike for a drive down to Bournemouth to learn English for the summer: a two-month sojourn in a bed and breakfast near the pier, with regular lessons from a tutor who would teach her class how to say 'ought' and help them master business English, a subcategory of the language that would vouchsafe future careers in the semi-conductor and textile industries of the Pearl River delta.

For his part, Mohammed was waiting for Chris's flight from San Francisco. The former, originally from Lahore, was at present based in Southall, while the latter, from Portland, Oregon, now lived in Silicon Valley – not that either man would attempt to discover these details about the other. In an otherwise uninhabited universe, how strange that one should so easily be able to sit in silence with another human being in a black Mercedes S-Class sedan. For both driver and passenger, the trip would be counted a success if the other party proved not to be a murderer or a thief. The hour and a half of stillness would be punctuated only by the occasional electronic command to turn left or right at the next junction, until the Mercedes reached a glass-fronted office building in Canary Wharf, where Chris was due to attend a meeting on the storage of financial data and Mohammed returned to the terminal to begin another journey, this time to Kent, with the no less mysterious or more talkative Mr K from Narita.

7 Many of the more conventional reunions seemed to beg the question of how their levels of excitement could be kept up. Maya had been waiting for this moment for the previous twelve hours. She had had butterflies since her plane crossed the coast of Ireland. At 9,000 metres up, she had anticipated Gianfranco's touch. But finally, after eight minutes of a sustained embrace, the couple had no alternative: it was time for them to go and find his car.

It seems curious but in the end appropriate that life should often put in our way, so near to the site of some of our most intense and heartfelt encounters, one of the greatest obstacles known to relationships: the requirement to pay for and then negotiate a way out of a multi-storey car park.

Then again, as we strain to remain civil under the unforgiving fluorescent lights, we may be reminded of one of the reasons we went travelling in the first place: to make sure that we would be better able to resist the mundane and angry moods in which daily life is so ready to embroil us.

The very brutality of the setting – the concrete floor marred with tyre marks and oil stains, the bays littered with abandoned trolleys and the ceilings echoing to the argumentative sounds of slamming doors and accelerating vehicles – encourages us to steel ourselves against a slide back into our worst possibilities. We may ask of our destinations, 'Help me to feel more generous, less afraid, always curious. Put a gap between me and my confusion; the whole of the Atlantic between me and my shame.' Travel agents would be wiser to ask us what we hope to change about our lives rather than simply where we wish to go.

The notion of the journey as a harbinger of resolution was once an essential element of the religious pilgrimage, defined as an excursion through the outer world undertaken in an effort to promote and reinforce an inner evolution. Christian theorists were not in the least troubled by the dangers, discomforts or expense posed by pilgrimages, for they regarded these and other apparent disadvantages as mechanisms whereby the

underlying spiritual intent of the trip could be rendered more vivid. Snowbound passes in the Alps, storms off the coast of Italy, brigands in Malta, corrupt Ottoman guards – all such trials merely helped to ensure that a trip would not be easily forgotten.

Whatever the benefits of prolific and convenient air travel, we may curse it for its smooth subversion of our attempts to use journeys to make lasting changes in our lives.

8 It was time to start packing up. In the connecting corridor leading to the Sofitel, I was intercepted by a fellow employee of the airport who was conducting a survey of newly arrived passengers, gathering their impressions of the terminal, from the signage to the lighting, the eating to the passport stamping. The responses were calibrated on a scale of 0 to 5, and the results would be tabulated as part of an internal review commissioned by the chief executive of Heathrow. I questioned the unusually protracted nature of this interview only in so far as it made me think of how seldom market researchers, with access to influential authorities, ask us to reflect on any of the more troubling issues we face in life more generally. On a scale of 0 to 5, how are we enjoying our marriages? Feeling about our careers? Dealing with the idea of one day dying?

I ordered my last commercially sponsored club sandwich in the hotel lounge. The planes were particularly loud as they passed overhead – so loud that as one MEA Airbus took off for Beirut, the waiter shouted, 'God help us!' in a way that startled me and my sole fellow diner, a businessman from Bangladesh en route to Canada.

I worried that I might never have another reason to leave the house. I felt how hard it is for writers to look beyond domestic experience. I dreamt of other possible residencies in institutions central to modern life – banks, nuclear power stations, governments, old people's homes – and of a kind of writing that could report on the world while still remaining irresponsible, subjective and a bit peculiar.

9 Just as passengers were concluding their journeys in the arrivals hall, above them, in departures, others were preparing to set off anew. BA138 from Mumbai was turning into BA295 to Chicago. Members of the crew were dispersing: the captain was driving to Hampshire, the chief purser was on a train to Bristol and the steward who had looked after the upper deck was already out of uniform (and humbled thereby, like a soldier without his regimental kit) and headed for a flat in Reading.

Travellers would soon start to forget their journeys. They would be back in the office, where they would have to compress a continent into a few sentences. They would have their first arguments with spouses and children. They would look at an English landscape and think nothing of it. They would forget the cicadas and the hopes they had conceived together on their last day in the Peloponnese.

But before long, they would start to grow curious once more about Dubrovnik and Prague, and regain their innocence with regard to the power of beaches and medieval streets. They would have fresh thoughts about renting a villa somewhere next year.

We forget everything: the books we read, the temples of Japan, the tombs of Luxor, the airline queues, our own foolishness. And so we gradually return to identifying happiness with elsewhere: twin rooms overlooking a harbour, a hilltop church boasting the remains of the Sicilian martyr St Agatha, a palm-fringed bungalow with complimentary evening buffet service. We recover an appetite for packing, hoping and screaming. We will need to go back and learn the important lessons of the airport all over again soon.

Acknowledgements

With many thanks to the following. At Mischief, Dan Glover (who had the idea), Charlotte Hutley and Seb Dilleyston. At BAA, Colin Matthews, Cat Jordan, Claire Lovelady and the Communications team, and Mike Brown and the Operations team. At Heathrow, Sofitel, British Airways, Gate Gourmet, the UK Border Authority and OCS. At Profile Books, Daniel Crewe, Ruth Killick and Paul Forty. Lesley Levene, Dorothy Straight and Fiona Screen for copy editing and proofreading. Richard Baker for the superlative images. Joana Niemeyer and David Pearson for the design. Caroline Dawnay and Nicole Aragi for the piloting. Charlotte, Samuel and Saul for another ruined August. In the text, some of the names have been changed to protect identities.

The first stage was by far the most important and the most difficult. Those in charge of the operation were inexperienced; moreover, they could not work to a coherent plan but had to adapt constantly to a military situation which the Red Army could not control. Thus, in Byelorussia, the evacuation operation took place between June and August, then was interrupted by the occupation of the whole of the republic. In the Ukraine, the transfer of the industries lasted from July to October; in the Baltic States it lasted until half-way through July, when the presence of German troops prevented the evacuation of men and goods from being carried out. The same was true of the Leningrad region in which the evacuation, begun in July, was interrupted in September by the beginning of the siege; it started up again at the end of 1942 and lasted for nearly a year, when it was the population which was evacuated. In the regions of Kalinin, Moscow, Voronetz, Tula, Rostov, Smolensk, Kursk and in the Crimea a limited evacuation began in the summer of 1941, then was accelerated between September and November. On 11 December 1941, the Evacuation Committee decreed that no more industry from the Moscow region would be moved. When it had become apparent that the German offensive before Moscow had failed, the problem arose of the return to the centre of the country (Moscow, Tula, Kalinin) of some enterprises. In June 1942 new German victories brought about a second wave of transfers, more limited than the first and better co-ordinated. The evacuation operations took place at this time mainly in the regions of Voroshilovgrad, Rostov, Stalingrad, Voronetz, Leningrad, Kerch and the Caucasus.

The factories and the workers were settled fairly quickly in reception centres. From early in July the factories began to be reassembled in the regions of Gorky, Kuibyshev, Saratov and along the Volga. In the middle of that month the first reassembling took place in the Urals and Siberia; from August 1941 to the spring of 1942 similar operations took place in Kazakhstan and Central Asia. In all the reception areas, bodies, set up at the local level, were given the task of resettling the people and the enterprises. Many contemporary eye-witness accounts tell of the difficulties, the incredible efforts and the dramas which permeated this titanic operation. These texts give a vivid picture of the dismantling of whole factories and the hasty departures, generally under enemy fire. The men who worked night and day were haunted by a single idea: would the Red Army hold out long enough? Often the operations were interrupted by the arrival of the Germans. Thus, the

Makeyevsky factory (Kirov), one of the oldest metallurgical enterprises in the Donbass, began to be dismantled on 10 October. On the 15th German troops arrived; 683 wagons loaded with material and 117 workers accompanied by their families had already left; in a few hours, the material which could not be dismantled was destroyed by the cadres who remained behind. The conditions of transfer of the workers and civilians, in general, were often terrible: overcrowded trains were held up for two to three weeks by innumcrable bottlenecks; on the waterways, the most unlikely craft were used to evacuate the population. In 1942 it was found necesssary to form, as well as the Evacuation Committee, a Central Office of Information to reply to all the requests for information about the displaced persons and the dispersed families. In February, a census of the evacuees was undertaken to enable families to be contacted and to bring a little order into the living conditions of the displaced persons. In order to integrate the 25 million displaced persons (17 million between June and October 1941 lodgings had to be found, often barracks and canteens and boarding schools for the children set up. Practically all the evacuated adults were found jobs in production; 1,530 enterprises were moved; and in the reception areas, the working population which in the country as a whole had decreased by 38 per cent between 1940 and 1942, in 1943 grew by 36 per cent in the Urals, 16 per cent on the Volga, 23 per cent in Siberia, 7 per cent in Kazakhstan and Central Asia. To be sure, this great effort did not prevent the Germans from taking possession during their advance of many factories which they either destroyed or used, but it did much to prevent the destruction of the Soviet industrial potential and enabled production, after substantial fall in 1941, to take off again the following year.

For the displaced population these years were very hard; lodged in very primitive conditions, often miles from the enterprise where they were employed, the evacuated labour force worked for more than ten hours a day in very harsh conditions without any respite. The recovery of production was very largely due to the almost superhuman effort of the civilians, especially of the women who almost everywhere took the place of the men who were called up.

Another weakness lay in the history of the officer corps. The purges had deprived the army of its experienced leaders and replaced them by young officers who, when war broke out, were joined by the survivors of the purges, taken out of the camps to be thrown on to the battlefield. The troops were thus led by men who were physically and morally broken and by untried officers, just out of the military academies, whose training in general had been far too rapid. At the beginning of the war, less than 10 per cent of the officers had finished higher military studies and 37 per cent had not finished their secondary military training. Seventy-five per cent of the officers and political commissars had only held their positions for a year. The war forged a coherent army, led by lucid and courageous commanders, but at first their heterogeneousness, their lack of experience and the recent events in the USSR made them very weak, compared with the German army which had been perfectly prepared for the coming struggle. Marshal Bagramiyan declared in 1945 that the destruction of the military leaders had been, in his opinion, one of the main causes of the defeats suffered by the USSR in the first year of the war.

The problems caused by the collapse of 1941 were not only military in kind; they were also above all political. The passage of an economically important part of Soviet territory and of 40 per cent of the population of the USSR under German control posed the problem of the cohesion of the USSR and, thus, of the regime itself. In the unoccupied territory a resistance movement had to be created and the authority of the political system which had been shaken by the defeats and by the situation created by the German occupation re-established. The attitude of 40 per cent of Soviet citizens towards Germany in the autumn of 1941 was an unknown factor which was to weigh heavily upon the future of the war in Russia and on the future history of the country. This is why the history of occupied Russia, and German policy in Russia, cannot be neglected.

GERMANY'S OSTPOLITIK

The invasion of the Soviet Union was not purely a military operation; it had a special function in Hitler's vision of Germany's future. Hitler's aims were precise: he intended to destroy Bolshevism, destroy the Soviet state and conquer the territory in which Germany would implant itself as a colonial power. The immediate interest – to prevent a Russian attack – and the economic objectives came a poor second to the wish to establish Germany hegemony in the East. This desire to dominate was justified in Hitler's eyes by a number of unshakable convictions. As far as the present was concerned, he thought that Bolshevism was specifically a Great Russian and Jewish phenomenon, two things equally detested by him. In the long term, he was convinced of the existence of an historical conflict, permanent and irremediable between the German and the Slav, a conflict which made Russia, irrespective of its political regime, a permanent danger to Germany. To free itself from this danger, Germany had to establish its domination over Russia once and for all. Lastly, he was convinced that the Slavs were an inferior race, unable to organise any form of State and argued from this that any form of political organisation in Russia must be suppressed for ever. Slavs did not have a State. These ideas were encouraged by Alfred Rosenberg, who was the architect of the *Ostpolitik*. In the summer of 1941, the problem of how to achieve the general objectives upon which, for the most part, all Hitler's collaborators were in agreement was resolved in several ways.

The character and ideas of Rosenberg are worth examining, because for twenty-five years he was Hitler's close collaborator, sharing his opinion with him. A Balt by birth and educated in Reval, Rosenberg, after the war, began to frequent National-Socialist circles where his book *The Myth of the XXth Century* earned him the reputation of a thinker. Simultaneously fascinated and repelled by Russia, Rosenberg, put forward more subtle arguments than those of Hitler as to how Russian power should be destroyed once and for all. For Rosenberg, Germany's enemy was not the conglomeration of peoples contained within the USSR, but only the Russian people, against whom he proposed to use the hostility of the other national groups. He based his arguments on the certainty that profound racial and cultural differences existed within the USSR and on the national grievances kept alive by the integrating policy of the Bolsheviks. Rosenberg suggested that the Russians should be brought back to the dimensions of Muscovy and isolated in the East by a *cordon sanitaire* formed of the Ukraine, Byelorussia, the Baltic States, the Caucasus and Central Asia. His final objectives was to destroy Russia and to root out its historic dynamism which, for centuries had pushed it towards the West, forcing it behind the glacis created by Germany to turn again towards the East.

From the summer of 1941, the *Ostpolitik* became a general concern of all the German leaders, and each claimed the right to express his opinion on the policy which should be followed in the occupied territories in order to impose German domination in the best way possible. Three ideas emerged, defining themselves in relation to Rosenberg's theses, who hitherto had been the only one to put forward plans of action. Rosenberg wanted Germany to exploit the national tensions, by giving the nations of the glacis State structures and governments, controlled by the Germans; in this way their national aspirations would be satisfied by associating them with Germany in the anti-Russian struggle, which would be waged without mercy and would include both the people and the regime. At the other extreme, Hitler, supported by the party apparatus represented by Bormann, opposed any concession to the peoples of the USSR, no matter who they were, and thought that Germany should fight the Soviet government and all the citizens of the USSR. In order to colonise, they had no need of allies, no matter how submissive they might be. Disagreeing with Hitler and against Rosenberg's selective policy, a quite different trend emerged among the administrators and the diplomatists, and for a time General Jodl succeeded in tempering Rosenberg's theses. Those who supported his plan thought that the Russian government must be dissociated from the people and that by winning the confidence of the people, the Soviet government and communist system could be destroyed. To play the nationalist card seemed to those who held this theory to be particularly dangerous since it involved wounding the nationalist feelings of half the population of the USSR, and throwing it back into the arms of their leaders. This is why they proposed to exploit the political and social grievances of the Soviet citizens and above all to foster the grievances of the peasants.

The first months of the war, marked by the lightning advance of the German army, confirmed Hitler in his opinion that it was unnecessary to seek the support of the population since victory was assured. This is why his ideas of violence and of the authoritarian organisation of the occupied peoples prevailed. The relatively favourable attitude of the first conquered people – in the Ukraine and Byelorussia – the capture of millions of men thrown into confusion by the conditions of the first battles, strengthened the theory that the east of Europe was populated by 'subhumans' (*Untermenschen*) who were accustomed to submitting to violence. At the same time, the extermination operations undertaken by the *Einsatzgruppen* (specialised groups of extermination squads set up in the spring of 1941) against the Jews, the communist leaders and others made terrible hecatombs among the civilian population of all ages, which brought in turn an upsurge of resistance, creating a vicious circle of reprisals. This again fed Hitler's conviction that violence alone was the way to communicate with the occupied peoples.

The organisation of the occupied territories reflected this refusal to dissociate the population from the regime. The occupied

Soviet territories were organised along the three broad lines: reincorporation into other States, civil administration and military administration. In the first group there were three territories: the region of Byalystok in western Byelorussia, seized by the USSR from Poland in 1939, and reincorporated into East Prussia in 1941; the western Ukraine, also seized from Poland reattached to the 'general government of Poland'; lastly Transnistria, lying between the Dnieper and the Bug and given to Romania. The civil administration covered two zones: the *Ostland* (the Baltic States and Byelorussia, minus the region lying to the east of the Berezina) and the Ukraine (up to Kharkov). The military administration kept under its authority a part of Byelorussia and the Ukraine and all the occupied territory of the republic of Russia, of the Crimea and of the Caucasus. Everywhere the higher or medium administrative power was in the hands of the occupying authorities. In the regions under civil administration the local authorities were only maintained at the level of the village or groups of villages (*volost*). In the cities, the local administration functioned under German control. In some cases, the military authorities tried to give the local administration wider powers, naturally under German control (in the Kuban, southern Caucasus, at Zhitomir). But these experiments were brief and isolated; generally, the local authorities were only permitted to function at the lower levels. In the zones lying near the front, no local administration was allowed to operate. Thus the occupation dismantled the Soviet administrative and legal system and placed the occupied peoples under the absolute authority of the German government bodies, civil or military. The decree of 17 August 1941 which set up the occupied territories, prefigured the transfer to the civil administration of the conquered regions if and when they were pacified; but no extension of the local powers was envisaged, even after the war had ended.

However, at the end of 1941 the *Blitzkrieg* was halted. Germany had to prepare for a long and hard war and was forced to rely on the economic and human resources of the East. From then on some degree of co-operation from the peoples became necessary, because the solution of Germany's economic problems depended largely on their good-will or hostility. From the end of 1941 it became clear that the attitude of the Russian population towards the Germans was neither as favourable nor as passive and submissive as had at first been supposed.

In the first weeks of their advance into Russia, the German troops had been greeted with a certain sympathy by the local population and Berlin had concluded that the Soviet population was ready to rise against its leaders and to receive the German army as a liberating army. In fact, this relatively favourable reception was influenced by the geographical conditions of the invasion. The regions first reached by the German armies were essentially parts of the Ukraine and of Byelorussia incorporated into the USSR in 1939, in which the Soviet regime had not had time to take root and in which the annexation of 1939–40 had been very unpopular. But as soon as the German army found itself faced by a population which had lived for a generation under the authority of the Soviet government the situation changed. In spite of the popular grievances, and the tragic memory of collectivisation and the purges, the population as a whole tended to see the German army as an enemy army. Moreover, after a few weeks, the harsh attitude and violence of the occupiers began to be known and helped to create the first group of partisans. The occupied population had to choose between support for the partisans and a pro-German attitude; there was no room for neutrality because of the many acts of violence of the German groups of intervention and of the army which hunted down the partisans. By the time the German leaders realised that there had to be a change in the *Ostpolitik*, based on co-operation with the occupied peoples, relations between the occupiers and the occupied had already deteriorated so badly that concessions were essential. These concessions, this temporary retreat from 'integral colonialisation' – and here Rosenberg's theory of the difference between ethnic peoples yielded to that of the social demands of the Soviet people – first affected agriculture.

Confronted by immense economic difficulties, Germany hoped that the conquest of the rich agricultural areas of the south would avert food shortages and, in the future, provide for German needs. The lines drawn by the invasion justified this optimism: 40 per cent of the cultivated surfaces, more than 100,000 kolkhozes, 3,000 tractor stations and 45 per cent of the livestock fell into the occupied zone. Of course, the terrible history of collectivisation and the recent invasions had exasperated the population of these regions. They hoped to gain from the German invasion only one thing, the opportunity to change the agrarian system. Just as it had at the end of the First World War, the collapse in the Second World War revealed starkly one of the constants of Russian society, the peasant's desire to own land. The agrarian question had been discussed in Berlin immediately after the invasion and several arguments had been advanced. A first school inspired by the colonial vision considered that the *status quo* must be maintained at all costs, as it had the merit of working fairly well, of enabling production to be easily controlled and of leaving no room for local initiative. Another possibility emerged from the situation created in the occupied territories in the summer of 1941. In many places, the peasants themselves solved the agrarian question and seized the collective land and the livestock. This is why the advocates of genuine collaboration with the Soviet population suggested that the situation should be accepted, that the people should be allowed to have their own way, hoping in return to win the support of the peasants which would show itself in an increased effort of production. Halfway between the two, many Soviet specialists advocated a planned transformation of agriculture, which would lead to the dissolution of the collectivist system. The two latter hypotheses from the beginning came up against the hostility of Goering who was in charge of the economic resources of Eastern Europe and who rejected them in the name of economic efficiency. He feared anything which could disorganise – even if only temporarily – agricultural production, and thought that the collective system was possibly the most suitable for the future needs of Germany, and that the exigencies of the present, like those of the future, justified the retention of the existing system. However, Goering agreed with his adversaries that the spontaneous action of the peasants was a problem and he tried to solve it by a formal concession. The expression 'kolkhoz' was to disappear and to be replaced by that of common farm *obshchii dvor*. This change of name did nothing to satisfy the peasant nor did it abate the quarrels envenomed by the military reports which showed on the one hand the desire of the peasants for a profound change and on the other their inertia which undermined the system of production. Gradually the argument shifted. In the beginning the argument of the *status quo* prevailed; in the autumn of 1941, the idea of a necessary change in the agrarian system had gained ground and its limits and modalities were being discussed. The halting of the German advance encouraged this line of argument. The resistance of Leningrad had shown that the populations of the cities were more determined to defend the political system than was the rural population. The idea of dissociating the peasants from the urban population – Rosenberg added a supplementary dimension by speaking of the possibility of making the Ukrainian peasantry the mainstay of this policy – seemed realistic and likely to weaken the resistance of the Soviet regime. The 'decollectivisers' argued that the concessions made in the occupied territory would raise the whole peasantry against the Soviet authorities and bring about an internal collapse of the regime. The memory of 1917 lay at the heart of all these arguments.

The plans of 1941 remained very limited, foreseeing a gradual evolution of the kolkhozes to communes in which the individual rights of the peasants would be recognised and in a future phase the return to private property. Against this project centred on the crucial problem of ownership of the land, which had on the whole the approval of the army, a counter-plan was presented by the Commissar for the Ukraine, Erich Koch, who was profoundly hostile from the beginning to decollectivisation. For Koch, it was not necessary to touch the economic structure in order to win over the peasants, because the same result could be obtained by minor political concessions, the recognition of indigenous representatives elected at village and district level. This plan, astonishing because it rested on a principle widely rejected by national socialism – the elective principle – although it was of no intrinsic interest, is worth noting because it was a sign of the concern of the German leaders about the hardening attitude of the population of the occupied territories. It also showed that at the end of 1941 after some local hesitation, the attitude of the population created a serious problem to the occupying authorities and thus destroyed the myth complacently spread in the years 1941–42 of a widespread collaboration.

The solution was found in an agrarian law promulgated on 26 February 1942 and published in all the occupied areas. The law abolished the whole of the Soviet agricultural legislation and organisation. It substituted for the kolkhoz communal ownership, (*obshchiinnoe khoziastvo*) which was seen as a transitional formula. Although the expression recalled the traditional Russian structures, the organisation was scarcely changed, except for the fact that the commune passed under the control of the German administration, which reserved for itself the right to dictate the volume of deliveries. The maintenance in practice of the collective system contained some minor compensations: the increase in the size of the individual plots, the outright ownership of the plots by the peasants and the reduction of taxes on this private property. In practice, these concessions were only partially implemented, the German authorities constantly putting a stop to the extension of the individual plots and using them as a spur to production and to obedience, which reduced the rights of the peasants to a minimum. This trend was aggravated by the fact that the peasants were discontented because the former system of paying wages, the *trudoden* (day-work), was maintained, and they reverted to producing for themselves. Beyond the communal ownership, no return to private property was envisaged, but there were to be agricultural co-operatives on the basis of village ownership controlled by the German administration (this was to be established after the war); the activities and objectives of the production would be set up by the administration according to a plan. The agrarian law mentioned incidentally the setting up of private farms which resulted from the reassembly of the plots, but this eventuality was very confused and vague and there did not ever seem to be any real prospect of this happening in the future. But until the transition of the phase of the co-operative had taken place, the authorities banned the appropriation of land, the distribution of which had always to be made under the authority of the administration.

As for the sovkhozes, as goods of the 'ex-Soviet State', they were confiscated by the German government which became their owners and were renamed State farms. Lastly, the German authorities tried to make the tractor centres the lynchpins of their political and economic control and they developed them wherever they were most needed, that is in the Ukraine.

The agrarian reform of 1942 concerned all the occupied territories and was to be extended to the whole of the USSR as the army advanced. In reality, there were several agrarian policies, founded on the competent authorities' assessment of the local conditions. Soviet collectivised agriculture thus gave way to very diversified systems, united only by the desire to exploit to the maximum the regions concerned.

In the Ukraine, the German agrarian reforms were only very moderately applied because the German authorities feared to upset the production of the regions which were richest in cereals, and also because they did not want to engage in a policy which ran counter to their post-war plans. The Ukraine was in the future to be open to German colonisation so why should a system of peasant ownership be introduced which would have to be speedily abolished? This was why, by the end of the occupation, less than half the land fit for cultivation had been given to the peasants in the structure of the co-operatives and no effort had been made to encourage private ownership. The Soviet system remained on the whole intact.

In Byelorussia and in the Caucasus, things were different. On the level of production, these regions were of less interest to Germany; furthermore, the nature of the soil lent itself less than did the Ukraine to cultivation of a communal type and more to small-holdings. For this reason the demands of the peasants were given more attention and the collectivised system set up after 1928 was largely dismantled. In Byelorussia, the reform took two aspects. The transition to co-operative farming was achieved and almost everywhere a communal form of ownership of the traditional type was re-established. At the same time, individual ownership grew perceptibly, sometimes reaching 4 to 5 ha. In 1943, the communes themselves were challenged and the trend was towards the breaking down of property which brought an increased effort by the peasants; production, in spite of the technical difficulties and scarcity of the labour force, was excellent.

Lastly, in the Caucasus, the agrarian policy from the first favoured the individual. In the rich regions of the Kuban the formula of communal ownership prevailed, accompanied by extended individual plots. In the mountainous regions in which a pastoral economy predominated, the German authorities passed without any transition to the formula of private ownership. In 1942, the kolkhozes were dissolved, the land distributed to the peasants and the livestock held in common to help to build up the herds. The mountain people of the Caucasus unreservedly approved this change in the structures of production.

Unequally applied, the agrarian reforms, however, did not have the desired effect on the Soviet peasants. The more time that went by, the more the peasants felt the burden of the occupation, the more they turned towards the partisans whom they aided to the best of their abilities. These growing reservations were explained not only by the undeniable patriotism, but also by the increasing pillage to which the German authorities subjected the Russian countryside. Most important was the economic pillage. Anxious to create in the East the conditions most likely to increase production, the Germans, in 1941, sent to the USSR machines and fertilisers which gave a slight impetus to local production. On the other hand, their levies were constantly increased and the completely arbitrary and unrealistic quotas imposed from Berlin reminded the peasants of the worst moments of the First Five-Year Plan. The reminder was all the more tragic in that, far more than under the authority of the Soviet planners, acts of sabotage (this was the term used to describe quotas which were not fulfilled) were ruthlessly punished. The law of scarcity, the contempt of the German authorities for the Slav peasant made his condition so harsh that all the psychological effects of the decollectivisation propaganda were cancelled out. Human beings were also treated as objects of pillage. The needs of the Reich's labour force led to large levies of workers, which exacerbated the feelings of the Soviet peasants and gave rise to grave problems in agriculture where there was already a shortage of labour. This swelled the ranks of the partisans and reforged the links between them and the peasants. From 1943 onwards Germany's agrarian policy found itself in an impasse. The reforms had undoubtedly weakened, or even dismantled the former structures, but they had not created a coherent new system able to satisfy the peasants nor, above all, had they achieved their first objective, of improving production. The production of cereals had fallen below that of the preceding year; the following year it was reduced by half, while at the same time the German levies doubled in 1943 and 1944, leaving to the local population an ever-diminishing share, a situation which could only help to widen the gulf between the occupiers and the peasants. From 1943 onwards, the Germans no longer had any coherent agrarian policy because they had realised that no concessions could win them the support of the peasants. From then on, violence took the place of an attempt at policy.

The reduction of the USSR to the level of an agricultural colony of Germany brought with it at the same time a reorganisation of agriculture and a dismantling of Soviet industry. The USSR had to supply Germany with minerals and oil, but it was not to retain any industrial structure which would enable it to convert its own raw materials. This was Hitler's initial plan, which was to reduce the USSR to the rank of a nation economically dependent on the great industrialised nations. However, the conditions of the war did not offer any opportunity to put this plan into operation. The dismantling of industry was the work of the Red Army which, in its retreat, carried off or destroyed the industrial potential of the USSR, so that the German troops found the territory which they occupied in the precise state to which they had been ordered to reduce it. Moreover, the technicians and the specialised workers had been evacuated as a matter of priority and the entire industrial activity was made unworkable because of this. The prolongation, unforeseen by the Germans, of the war in the East was to modify still further the realities of the industrial problem. The needs of a huge army, far away from its vital centres, demanded the reconstruction of a minimal Soviet industrial infrastructure and this led the Germans, contrary to all their forecasts, to revive the Soviet industry that they had wanted to destroy. The mines of the Donbass, various other mineral deposits and the oilfields of the Caucasus had to be put back into operation – all had been so badly sabotaged that it took two years of effort before there was more than a token production. The Dnieprostroy dam, the steel works of Zaporozhiye and many other industries were gradually brought back into operation, but everywhere production remained below normal and the effort, which was crowned with success in 1943 simply meant that it was easier for the Russians to recover after the war. One sector was of particular concern to the Germans, that of the construction of agricultural equipment; one such industry was created from scratch, in the Ukraine between 1941 and 1943. Thus the initial intentions of the Germans in practice came to nothing, as, driven by necessity, they rebuilt as far as possible Soviet industrial potential. The partially reconstructed industry was confiscated by the German government which considered itself the legitimate heir to the goods of the Soviet State. On the industrial plane, the idea of a return to the system of private ownership was never discussed, with the exception of the case of the Baltic States which were given a privileged regime. In Estonia, Lithuania and Latvia, the principle of the return to private ownership was admitted and some transfers of industrial property were effected. But this was an exception, running counter to the general policy on industrial property. This trend showed the German distrust of the urban population of the USSR. Although the Germans had tried to satisfy the demands of the peasants, they had from the first agreed that the working class had been bolshevised and that generally the urban population must be held on a tight rein and forced to comply with the needs of production. The lot of the working class was, during that period, particularly hard because the cities were very badly provisioned, the prices high and the wages reduced as far as possible in order to reduce the cost of production. Only one social class was given different treatment, that of the craftsmen. The German authorities tried to revive the craftsmen within a corporatist system. This policy corresponded with the overall ideological trend of national socialism, which was attached to the traditional system of labour relations, but mainly it was an attempt to divide the urban population, to raise up against the hostile working class a class of privileged craftsmen and by so doing it was hoped to win them over to the New Order. But these attempts to organise labour on different bases yielded only very poor results.

German economic policy in the end came up against the same fundamental difficulties as had the Soviet government; it could not conciliate the peasants. Neither in the country nor in the towns did the Soviet population want to co-operate with the Germans' attempt to reduce the country to slavery. The advocates of the policy of relying on the support of the whole population against the Soviet authorities were forced to concede that their policy had failed. After that, the German concessions were directed to other realms, above all towards another divisive element, the nationalities; Rosenberg's ideas seemed then to have won.

The nationality policy of Germany was a perpetual compromise between general principles and individual visions, even passions. The general principles were the desire to isolate the Russians behind a glacis of non-Russian nations, placed at the bottom of the scale of nations from the ethnically different groups. This principle posed many questions: in the first place, that of the future of the nations concerned. Was the aim of the differentiation to give them a political status which recognised their personality and their rights? Or were they simply intended to weaken Russia? Should all the non-Russian nations be put on the same footing? Or should there also be a scale here? On the first point, the hostility was between Rosenberg, who advocated a genuine national policy, recognising that the national groups had the right to their own existence, and Hitler, who, although he wanted to divide the nations of the USSR, refused to set up in the East any national States at all; he held all the inhabitants in equal contempt. The only ones who found favour in his eyes were the Muslims because of the attitude adopted by the Caucasians towards Germany in 1942. But in general, Hitler was hostile to the existence of any sovereign State in the East and therefore condemned any form of local power which, he saw as a first step towards sovereignty. On this point, Hitler's theories were to leave a decisive mark on the policy towards the nationalities of the years 1941–44. The other unknown factor was the fate of each nation. Here genuine differences were to exist until 1944, linked to the economic and geographic position of the nations, to their reactions towards Germany and to their historical background. Several situations could be discerned, that of the 'privileged' Baltic, Caucasian and Muslim States and that of the Ukraine which had a central place in the German plans.

The Baltic States, for obvious historical reasons, had a privileged place in the German conceptions. The Baltic world was, above all, a German advance post towards the East, partially populated until 1939 by Germans and in which German culture had always been respected. Moreover, the Balts were not Slavs: they had played a leading role in Russia and as regards the Russians they had a superiority complex which, since 1939, had been more than tinged with hostility. Compared with Asia, the Baltic world represented Western civilisation and it had been the centre of the Russian thrusts towards the West and of Western penetration of Russia. From this situation the Baltic world held its letters patent which, in the political recasting which was taking place, made it worthy to be associated with the destiny of Germany. However, the terms of this association were not particularly favourable to the national development of the Balts. Their future ran along two equally bitter lines, the influx of a German population and the setting up, at best, of a protectorate. In fact, the Germans had only a limited confidence in the Balts, in spite of the profound differences between them and the Slavs. Although the Estonians were considered sufficiently Germanised to be 'salvaged', the Latvians were held to be very Russified; as for the Lithuanians, they were above all 'Jewified' according to the German experts, which meant that a large part of the population of Latvia and Lithuania was to be deported to the East and replaced by German settlers. Finally, the privileged treatment applied more to the Baltic lands than to the peoples who were historically linked to them and for whom there was to be little room in the future organisations. In the immediate future, however, the Baltic peoples benefited from a special status, in the form of national governments, pro-German to be sure, but which to some extent acted as a buffer between the population and the occupying forces, and which benefited from some degree of support from the population. Thus, while elsewhere intellectual life was very restricted, in the Baltic States on the contrary it could continue normally. The local authorities, emboldened by the concessions made to them, increasingly defended the general desire for genuine autonomy and so aroused Hitler's hostility. At the same time, when the conflict erupted in 1943, Germany was desperately in need of volunteers in the occupied regions, and thus had to take account of the aspirations of the people. A compromise solution was reached at the beginning of 1944, in which the German concessions, restricted to the cultural sphere, gave the local authorities some openings. The question of political autonomy remained unresolved up to the very end. In any case, at this time, the anti-German attitude of the Balts had increased to such an extent that any political concessions had become dangerous. The occupation of the Baltic States was on the whole less brutal than elsewhere, the concessions made to the national aspirations more extensive, but these advantages were only relative and fell very far short of the popular will. It was not safe to play the nationalist card and then to refuse any outlet for the expectations which had been aroused. The German occupation had consolidated the national sentiments of the people already very attached to their historic tradition and had shown them that they could not free themselves from the grip of the Soviet Union by relying upon Germany.

Very different factors inspired the policy in the Caucasus and in the Muslim countries. In the Caucasus – that is, in the three republics of Georgia, Armenia and Azerbaizhan and in the republics of the autonomous regions – German sights were set on the oil wealth and not on a policy of colonisation. The ideas of Hitler and of his collaborators about the population of the Caucasus were determined simultaneously by the history and by the particular influences. The Georgian resistance to Sovietisation, the existence of the hill tribes who were reputed to be very independent, the memory of the independent States of 1920 which had been recognised internationally and the absence of Slavs all inclined Hitler to regard the Caucasian peoples as Aryans of a special kind. Moreover, a large Caucasian emigration established in Germany since the 1920s had been particularly active and had helped to develop among the German theorists of the *Ostpolitik* the idea of staking everything on the break-up of the USSR. The Germans thought of the Caucasus not as a glacis intended to contain the Russians but above all as a platform for an eventual advance towards the oil-producing regions of Iraq and Iran; in the meantime, the Germans wanted to avoid any disorder and tried to conciliate the Caucasus rather than transform it.

The ethnic complexity of the region, and the cultural and religious differences between the peoples who inhabited it could have gravely embarrassed German policy had the occupation been prolonged; but it was relatively short and did not affect the main nationalities of the region, which enabled the occupiers to avoid having to arbitrate in local conflicts. The overall plan concerning the organisation of the Caucasus was based on the idea of giving some political autonomy to the nationalities concerned and of placing this autonomy within the framework of a German protectorate. This was rendered necessary by the national rivalries which would inevitably find expression. Even before the German advance towards the Caucasus, the *émigrés* in Berlin were trying to influence the decisions on the future of the Caucasus and were quarrelling furiously over the three arguments which the abstract strategists hotly supported: a Caucasian federation, with Georgia as the pivot, a plan which alarmed both the Armenian and the Turkish peoples; a Caucasian federation with a Turko-Muslim pivot which alarmed the Armenians and the Georgians, who were for once in agreement; lastly a juxtaposition of autonomous States which the German protectorate would compel to co-exist peacefully. While in the Berlin offices the map of the Caucasus was being redrawn, on the spot the realities of the problem were emerging. The German government did not want to offend Turkey which was still neutral, but concerned about the fate of the Caucasian Turks and traditionally hostile to Georgia. Furthermore, the German armies, impressed by the attitude of the Caucasian hill folk and thinking that for the first time they had found genuine allies in the USSR, drew up for them a liberal policy which left room for native initiative. The ambitious plans of the overall organisation of the Caucasus were destroyed by local realities.

In the summer of 1942, in fact, when the German army reached the shores of the Black Sea, it was faced in the mountains separating the Caspian from the Black Sea with a new situation – one which it had encountered but in more vague terms in the Ukraine in 1941. The mountain peoples of the Caucasus had reacted to the Soviet military collapse by rebelling against Moscow under cover of the retreat of the Red Army. These risings had eliminated the Soviet authorities and they had been replaced by the local authorities, decollectivisation and the re-establishment of the national, religious and cultural traditions. This movement was particularly strong among the Karachais where Madzhir Koshkarev seized power in the capital and welcomed the German troops, and among the Kabards and the Balkars. The concessions made to these people by the German army – which was the master of the situation in the Caucasus and in which neither the civilian authorities nor the SS could intervene – were considerable and created a special system. The army recognised the authorities of the Karachai national governmental committees (led by a peasant Kadi Bairamukov) and Kabardo-Balkar (led by Selim Shadov). These local governments, supported by the population, enjoyed a genuine autonomy in cultural and religious affairs and even carried some political responsibilities, above all in the Karachais region. There, contrary to the rule established in the rest of the USSR, the German army did not confiscate the goods of the Soviet State, but handed them over to the regional authorities who were thus made the beneficiaries of the former government. The decollectivisation launched by the inhabitants themselves was approved by the German army and could be put into effect unimpeded. During the months of the occupation, the mountain people were free to renew their religious practices and often the German army took part in the traditional ceremonies. Even anti-Semitism was not so virulent in the Caucasus. In fact, the national governments were opposed to the effect of racial discrimination upon the small groups (25,800 in the 1926 census) of Jews among the mountain tribes (*Daghshufut* or in Russian *Gorskiye-evreyi*) who had arrived during the Sassanid epoch and were completely integrated in local life. However, in 1943, because of growing military difficulties, relations between the local population and the German army deteriorated and by the time the army retreated, the national governmental committees were already partially discredited. The fate of another Muslim people – Tatars of the Crimea – was passionately argued after the invasion of the Crimea which Germany intended to colonise extensively. There Slavs and Turks co-existed, and in the German plans one could see just how deeply the 'Muslim myth' linked to the Caucasian affair had suddenly taken root in Berlin. While the German plans included the deportation to Russia of all the Slavs in order to make room for future settlers, the Crimean Tatars protected by Turkish fellow feeling were able in 1942 to set up a Muslim Central Committee with its headquarters in Simferopol.

However, the German authorities were not inclined to move as quickly in the Crimea as in the Caucasus, because they feared the rebirth under their aegis of the Panturanian movement of the 1920s which the Soviet government had successfully broken. This disquiet was well founded, because the former leaders of the Tatar national movement became active once more, especially one of the leaders of the *milli firka*, Ahmed Ozenbashly, who tried to set up a representative assembly or *kurultay*. These anxieties coincided with the advance of the Red Army which in November 1943 had arrived at the frontiers of the Crimea which it reconquered in the spring of 1944 with the fall of Perekop and of Kerch.

The main effect of German policy in the Crimea was finally to help towards the resurrection of the Tatar national organisations and to give a powerful impetus to the nationalism which had been prevented from expressing itself since 1923. In the Crimea, as in the Caucasus, Germany's fundamental hostility to Russia had led her to abandon her racism in order to place in the Muslim peoples and the Kalmuks a confidence as great as the confidence they had placed in the Balts, who were, however, far closer to them.

The fate of the two other occupied regions – the Ukraine and

Byelorussia – was on the whole very different. The Ukraine had already found a place in German political calculations at the end of the first World War and the idea of weakening Russia or Poland by fostering Ukrainian nationalism had been approved in 1919 and 1939. After the invasion of the USSR, the Ukraine occupied a special place in the German plans because of its geographical position – a buffer State and at the same time a bridgehead for expansion towards the south and its natural riches.

However, there were two opposing arguments about the future of the Ukraine. Hitler was completely indifferent to Ukrainian aspirations and thought that the territory should be organised immediately for its future use and that the population who inhabited it should be disregarded completely. The master-minds of the *Ostpolitik*, on the contrary, thought that the Ukraine could only play the part which had been assigned to it as regards Russia if – at first at least – Ukrainian national aspirations were recognised in their entirety. This trend, recommended by the *émigrés* whose ranks were swollen by the Ukrainian nationalists liberated from the Polish prisons in 1939 – such as Bandera who was to play an important part even after 1945 – urged separatism and the setting up of an independent national Ukrainian State.

The invasion of the Ukraine revealed that the plans drawn up in Berlin were not always suited to the local situation. Although the *émigrés* were fiercely separatist, the local population were far more interested in the problem of the land than of separatism. As for the intelligentsia, it was aware of the German contempt for the Slavs and that the ultimate destiny of the Ukraine was to become a German colony. In the meantime, although during the first months the Ukrainian population remained in a state of wary expectation, the German land policy turned it very quickly away from separatism. Even in the Ukraine, from the first, Hitler's argument prevailed. The Ukrainians, after all were only Slavs and their physical vitality, their fertility revealed by the large families in the country, the national tendencies given prominence by Rosenberg, were all considered as the source of delicate problems in the future. From the beginning, the Ukrainians had to be subjugated and not encouraged; this was the reason why in 1941 German policy was marked by the brutalities, the ill-treatment (the corporal punishments astounded the Ukrainians) and the desire to degrade this people economically, physically and morally. The local intelligentsia, suspected of spreading nationalism, was decimated, the national culture and traditions persecuted. The Ukrainians, traditionally anti-

Semitic, nevertheless reacted indignantly to the massacre of the Jews in the Ukraine and pinned all their hopes on the partisans and the Red Army. In 1942 the Germans who had been welcomed with relief six months earlier had been totally discredited.

The same thing happened in Byelorussia. Like the Ukraine, Byelorussia in 1941 was difficult to define because of the territorial additions of 1939 which had changed its population and had included in the republic elements which were barely Soviet and marked by a very strong national spirit. At the same time, Byelorussia, which was economically poor did not present the same interest to Germany as did the Ukraine; this is why this territory which had already a high proportion of Jewish inhabitants was transformed into a place of transit for the Western Jews awaiting the 'final solution'. The result of this policy, applied from 1941, was that Byelorussia became a place where the Jews were massacred – massacres in which the horrified local population was forced to take part. Byelorussia, which the army regarded with an indulgent eye because it was poor and underequipped intellectually, and thus unlikely to put up any real resistance to the German domination, rapidly became a centre of resistance; the geographic conditions were favourable to the growth of the partisans. The reprisals and the growing exacerbation of the local population all strengthened the resistance. In 1943, uneasy about the evolution of the Byelorussians, some of whom had welcomed their arrival, the German authorities decided to modify their policy and try conciliation. On 21 December 1943 a 'Byelorussian Central Council' was set up, as a pledge of the government which the population had been promised. But it was too late; German defeat was now a certainty and the phantom government established by the occupiers received no support.

German policy towards the nationalities seemed, when it is examined region by region, to have been very incoherent. It is, however, possible to discern its guiding lines. Wherever German economic interests, wherever the great colonial plan were at stake, Hitler refused to listen to the national theories of Rosenberg; he wanted from the beginning to create the conditions for a future Germanisation of the territories and this explains the ruthless character of the occupation in the Ukraine. Wherever, on the contrary, the economic interests were less important or non-existent, wherever there were other preoccupations – foreign policy or matters of strategy – the line followed was less clear and sometimes yielded to local realities. However, and Rosenberg had insisted on this, if a national disintegration of the USSR could be encouraged it was above all in the Ukraine that this policy should be applied. The size of the Ukrainian population made a positive policy in this respect far more feasible than the concessions made to the isolated groups in the mountains and in any case far less integrated within the USSR. Had the Ukraine swung over to the Germans taking Byelorussia with it, there would have been no partisans and the rear of the German army would have gained substantial support. Rosenberg's speculations were not completely without substance because the population of the Ukraine and of Byelorussia did waver for a brief but decisive moment.

By 1943, the situation had finally clarified and the population had united against Germany; no gesture of appeasement, no concession, no matter how important, could turn the occupied peoples from the pro-Soviet attitude which had emerged from the long months of suffering and terror. From then on for the weakened Germany the objectives of the 'national policy' changed completely. Germany needed workers for its factories, mercenaries for its armies. It was difficult to find workers, for the able-bodied had joined the ranks of the partisans in order to evade the requisitions. The Germans tried to attract soldiers by setting up national units, even including the Russians. In the first months of the war, German propaganda was designed to persuade the soldiers of the Red Army to desert and to join the ranks of the German army in order to 'liberate' their country; by 1943, these efforts reached only the prisoners and the population of the occupied territories. But here Germany's national policy found itself at a dead end. Germany could not hope to recruit soldiers simultaneously among the Russians and the non-Russians, who were asked to separate from Russia. This contradiction was never solved, and the Vlasov movement and its army set up by a general of the Red Army who had gone over to Germany after being taken prisoner in July 1942, co-existed with the autochthonous battalions (Turks, Caucasians, Cossacks, etc.). From among the national corps, that of Vlasov was the most important, in that its leader's aim was to raise the army against the government, in order to reconstruct the Soviet State on new democratic but not capitalist foundations.

Vlasov's propaganda campaign from 1943 onwards forced Moscow to redouble its efforts in order to prevent the Red Army from heeding his proposals. However, it was not the Soviet government which deprived Vlasov of all credibility but the Germans themselves. From the first they undermined his position by supporting the idea of a Russia reduced to its simplest territorial expression, and by refusing to clarify their vision of the political future of the Russian State. The increase of the non-Russian regiments confirmed their determination to take from Russia all the non-Russian territories. Lastly, during the last months of the war, the Russian and non-Russian battalions were thrown into battle on the Western front, although they had been raised as troops of national liberation. Thus the intention of German policy was revealed: the men whom it had managed to draw over to its side in order to speed up the disintegration of the Soviet State were for them purely political tools and mercenaries, and not the pillars of a future organisation of the USSR. The Kremlin had no need to inveigh against these men: in the minds of the people they formed tragic cohorts of the collaborators from every nation who in 1945 were the victims of the collapse of Germany.

The German leaders were convinced in 1941 that the Soviet population, exacerbated by its own regime, were ready to accept German domination and sometimes this belief was justified by the facts. The aim of the *Ostpolitik* was to destroy, by various means, the USSR and the structures of the socialist system in order to smooth the path towards the German hegemony. This plan, which was basically designed to further German interests only, entailed the application of many policies which the conditions of war constantly diluted and distorted. Although the effects of the *Ostpolitik* for Germany were in the end negative – the population of the occupied territories was raised against Germany by the brutalities but also by the broken promises – they were considerable for the Soviet regime in the course of the war and afterwards. In the course of the war, the great characteristics of German policy in the occupied territory, added to the centrifugal tendencies provoked by the occupation, were to contribute in all likelihood to orientating Stalin towards an internal political revision in various fields.

THE TRANSFORMATION OF THE GOVERNMENT AND OF STALIN'S POLICY

The German invasion (this is confirmed by many testimonies) caught Stalin unawares. In a few hours his political and strategic ideas were swept away and the whole system seemed to be tottering. The speed of the German advance and the inability of the Red Army to escape from the invader all pointed to false calculations and helped to increase the disarray of the entire USSR. For several days, Stalin withdrew into himself and his dismayed collaborators were unable to get from him either directives or any reaction or even signs of interest. The USSR no longer had a leader. Officials of the Comintern, at that time working in Moscow, have reported the quasi-funereal atmosphere in the Kremlin. There was a panic too because all the leaders thought, judging from the reactions registered that the national unity of the USSR had been destroyed. While the people waited for Stalin to give some sign that the USSR had not been completely knocked out, it was Molotov, the Minister of Foreign Affairs, who on 22 June broadcast a short and unemphatic speech, as Stalin, having told his collaborators that, 'What Lenin created we have lost forever', seemed to have divorced himself from the situation. The first bodies set up to respond to the invasion – the Great General Headquarters, formed on 23 June, the National Defence Committee (GKO) created on 30 June – either met without him or if he were present he was distant, passive and silent. The GKO which assumed full powers, was from the beginning collegial in character. As well as Stalin, Molotov, Malenkov, Beriya and Voroshilov were members: they were later joined by Mikoyan and Kaganovich. These were the darkest days of the war because it was then that the Soviet State tottered. The disarray of the leaders affected the territories into which the Germans had penetrated: no one knew which authority they should obey. The Party retreated into clandestinity; the State was invisible and this helped from the first to strengthen the position of the occupiers who installed themselves in the sudden vacuum of power.

On 3 July, finally, Stalin pulled himself together and the defence was begun. It was clear to what extent the passage of time had marked the political life of the USSR. Stalin's silence brought all life and reaction completely to a standstill: his reappearance was the symbol of the national revival. Rarely had power been so personified, so intimately linked to one man; rarely had the collaborators of a leader seemed so like mere onlookers. The ten days which elapsed between the invasion and Stalin's reappearance gave proof of how total Stalin's power was.

His reaction, when it came, was remarkable because it showed awareness of the situation, took account of the state of mind of the Soviet people, of the doubts which weighed upon the regime, of the revived grievances and of the hesitations in fact of the invasion. The man who addressed the USSR on 3 July was not the leader of a Party who understood only a minority of the population; nor was he the leader of a socialist State, for the socialist transformation had left behind a residue of great resentment; he was the head of a *nation* the diverse components of which had an heroic past, which had known how to resist many invaders, of a nation linked to this historic past and its native soil. The solidarity which Stalin called for was not one of ideas, it was that of the past. He left out of his speech the greeting traditional in the USSR 'Comrades', which evoked the links of class, and found again the old vocabulary of the leaders of the Russian State, the vocabulary which bound together a national community throughout the ages 'Brothers and sisters, a great danger threatens our country'. This was not a question of the regime nor of the Party, but of that more remote element of unity which for some years had rediscovered its rightful place: the Motherland. The movement of national renaissance which had been glimpsed before 1939 unfurled during the war and in an ever-growing way the Russian heritage, the whole history of Russia was called to the support of the new patriotism.

On 10 November for the anniversary of the Revolution, in the underground station Mayakovsky, Stalin appealed to the Soviet people to resist the invader by rallying around Russian patriotism. The foundation of this patriotism was the 'great Russian nation of Plekhanov, of Lenin, of Byelinsky, of Chernichevsky, of Pushkin, of Tolstoy, of Glinka and of Pavlov, of Suvorov and Kutuzov'. The next day in Red Square, reviewing the troops who were leaving for the front, he besought them to resist by evoking the pleiad of heroes of Russian history: 'May you be inspired by the glorious example of our ancestors, Alexander Nevsky, Dimitri Donskoy, Kuzma Minin, Dimitri Pozharski, Alexander Suvorov, Michael Kutuzov'. In 1942 *Pravda* published a series of articles commemorating these heroes. The historical chain, at one time broken, was being renewed and the sovereigns of the old Russia who had fought to shake off the Tatar yoke took precedence over the heroes of the Civil War. Throughout the war, this hierarchy of the historical 'fathers' of Russia was to be constantly evoked and in his office Stalin contemplated simultaneously the portrait of Lenin and those of the military leaders who had fought against Napoleon.

By the autumn of 1941, the Soviet nation had found its final content. The traditionalist definition of the whole community, which down the centuries united all the children of the same land over and above the interests of particular groups, had taken the place of the socially selective definition of the Marxists. The whole Stalinist vocabulary shifted in a national direction. Against the invader it was not the proletariat whom he called upon but the people 'to rise as one' (*narodnoye opolchenye*); and the war was to be called from 1943 onwards 'the Great Patriotic War' (*velikaya otechest-vennaya voyna*). The new vision implied by this national terminology did not concern Russia alone: it contained a new definition of the relations between the nations and the loyalties of the USSR. If in the past Soviet ideology had opposed the USSR to the capitalist nations, the notion of the *democratic camp*, applied to the USSR, to Great Britain and to the United States, substituted anti-Nazi solidarity for class solidarity. The war against Germany became a 'just war', a war of liberation and Stalin dissociated two forms of national sentiment: the positive one of the Soviet people because its objective goal was only to strengthen and save the national community, the other, negative, because offensive, of Germany which was in reality *imperialism* and not nationalism. Of course it may be objected that these definitions sprang from tactical necessities and that they cannot be assimilated to a radical ideological break.

However, they were confirmed by the concomitant appearance of a new type of solidarity drawn also from the past history of Russia, *Slavism*, the rebirth of which was one of the fundamental aspects of Stalinist war ideology. Even at the XVIIth Party Congress in 1934 Stalin had stressed that the German threat was aimed in the first place at the Slav peoples. Although during the years between 1935 and 1938 Stalinist diplomacy was directed towards the non-Slav peoples of Western Europe, then between 1939 and 1941 towards Germany to the detriment of the Slavs in Poland, this break in the Slav orientation was clearly determined by the exigencies of defence. However, the speech of 6 November 1941 which spoke of the Russian people, reintroduced at the same time the notion of the solidarity of the Slav peoples which the Committee of All Slavs created in 1941 with Major Gundorov as President, turned into reality. This defensive Slavism of the war years was to be developed after the victory in a more positive way, which was to be one of the foundations of the post-war East European order. On 12 December 1943 during the signature of the Soviet-Czechoslovak Treaty, Stalin said that all the Slav peoples should unite (in order to prevent any repetition of German expansionism), and he developed this theme on the occasion of the signature of the Soviet-Polish Treaty on 21 April 1945. From past experience, Stalin drew a picture of the future which raised the tactic of the moment to the level of a permanent principle. To be sure, later Slavism was no longer to be affirmed in the same way, but it was its place in the chapter of the changes which took place in the war ideology. Above all, Slavism, like the resurrection of the national, essentially Russian, principle, and the revision of the foundations of the relations between the nations, was translated into an evolution which was already included in the Constitution of 1936. The USSR was the established State of a defined nation and not only a revolutionary base which was destined to refashion itself constantly as a result of the expansion of the Revolution. The stabilisation of the USSR, according to the dissociation of the Soviet State and the world revolution begun long ago, were processes to which the Second World War gave a decisive turn. In May 1943 Stalin dissolved the Communist International because it had 'achieved its historical mission'. Of course it can be objected that this was a concession to Russia's allies designed to prove to them that the USSR was no longer a threat to the international order. In reality this was not only opportunistic; Stalin had already envisaged this suppression at the beginning of the war but he had not wanted to react under Nazi pressure nor from a position of weakness. He acted after Stalingrad when the Soviet army had begun its liberating advance towards the Caucasus and into the Ukraine and was giving proof of the rediscovered strength of the Soviet State.

Once the International had been dissolved, the USSR drew ever further away from the world outlook of the young Bolshevik State in order to strengthen the national character of the Soviet State. The old revolutionary song the 'Internationale' which had been the anthem of the USSR since 1917 was replaced by a patriotic anthem in 1943–44.

Even in 1942 this change could be foreseen. The traditional slogan of the Soviet newspapers, 'Proletariat of the world unite' was replaced by the patriotic appeal, 'Fight the fascist invader'. The changes which took place in the army were also signs of the renewed link with the national pre-revolutionary past. The epaulettes torn off during the Revolution as symbols of oppression appeared again to decorate the military uniforms, the formations of Cossacks, spearheads against the Bolsheviks, reappeared and the regiments adopted again names traditionally honoured in the Russian army.

Stalin created in November 1942 the orders of Suvorov (he awarded himself the order of Surorov, first class, on 6 November 1943), of Kutuzov and of Alexander Nevsky. The functions of the political commissars of the Red Army were suppressed in November 1942 and after the middle of 1943 the promotion of officers to the rank of general greatly increased. 'Surorov' schools were established to train officer cadets.

The restoration of the Russian nation posed the problem of the policy to be adopted towards the non-Russian peoples, and this was all the more urgent because the German occupation had shown the responsiveness of the non-Russian peoples to their national sentiments. Here again, the Stalinist re-evaluation was considerable. While during the pre-war years the revival of the national past of the non-Russian peoples had been very cautious and restricted, in the war years their past was magnified. The journal *Propagandist* in 1942 deplored

the almost total lack of books on the national heroes, on the participation and military co-operation of the different peoples of our country in the patriotic people's wars against foreign conquerors, on the love of the motherland and the love of liberty of these peoples. There exists among the peoples of our motherland a burning desire to know more about the heroism of their ancestors.

This text gives the measure of the concessions made by the Soviet regime to the national sentiments of the non-Russians. Only a patriotic upsurge could have saved the USSR in 1941. Russian patriotism, however, was not the universal foundation of this patriotism; it could not be extended among the other peoples of the USSR. However, even in the desperate months which followed the invasion, the regime always remained aware of the limits which it could assign to the development of the non-Russian nationalisms. This acute awareness of the ambiguity of nationalism made it necessary to work out a complex policy. On the one hand, the press constantly insisted on the historic heroism of the non-Russian peoples and films were produced and books published relating the remarkable episodes in their past history. At the same time, it is interesting to see that the non-Russian heroes were rarely exalted. In the innumerable series of pamphlets published during the war years in Russian in millions of copies, the aim of which was to popularise the heroes of the pre-revolutionary history, the non-Russian heroes represented barely 5 per cent of the publications. The only two people who emerged from this silence were Iman Shamil, hero of the resistance in the Caucasus, and Bogdan Hmelnitsky. In both cases, the men praised were part of the historic heritage of the occupied peoples and Hmelnitsky was exalted by the propaganda as the man who had united the Ukraine to Russia. Here we come to an essential aspect of the ideology put before the non-

Russian peoples during the war. The emphasis was laid on what united them to Russia or separated them from the neighbouring peoples and justified the union with Russia. It is in this perspective that, together with the military orders created by Stalin which referred to the Russian heroes, was added the order of Bogdan Hmelnitsky, symbol of the unity and the equality of the peoples of the USSR. More modest publications also appeared in the national languages. Books appeared in Georgia glorifying Queen Tamar and David the Builder who fought against the Turks and the King, George V, who drove the Tatars from his country. The Armenians were invited to glorify David Bek, who in the eighteenth century led a revolt against the Persians; the Azerbaizhanis, Babek, hero of the anti-Arab resistance. In the Ukraine, after its liberation, the figure of Daniel of Galicia, hero of the thirteenth century, was added to that of Bogdan Hmelnitsky. As for Central Asia, the history of which had been a long series of battles against Russian advances, its heroes seemed to have been forgotten in this recalling of the past. What was put forward, together with the past of the struggle against the foreign invaders, generally common enemies of Russia and the other peoples of the USSR (Turks, Tatars, etc.), was precisely the old community of interests and of destinies. Here again the past was projected upon the present and the conclusion could be drawn that past history had forged a community of equal peoples, the peoples of the USSR among which the Russian people – *primus inter pares* – assumed the essential task of the common defence in the war. In the thematic ideology of the relations between the nations during the war a new idea gradually appeared, that of the commuunity of equals, with one more responsible and more powerful people. In 1917 the Bolsheviks had rejected the paternalistic idea of an 'elder brother' of the peoples of the USSR; it began to reappear in a diluted form during the war years.

However, ideology was not the only sphere in which there was a revision of the national policy. In practice, Stalin relaxed the controls which weighed on the non-Russian nations and once again in the governmental apparatuses and the national parties indigenous cadres began to appear. This evolution was made necessary by the war effort which affected especially the Russians and made it necessary for an appeal to be made to the non-Russian cadres in the rear. These cadres favoured a development of national cultures and traditions and so the war years were favourable to a certain renaissance of the nations of the USSR.

This relative liberalism was particularly apparent in the evolution of the religious policy of the USSR which helped to restore to all the peoples, including the Russians, a still very important aspect of their culture and of their historical heritage. Here again Stalin's realistic policy responded to the oscillations of the German *Ostpolitik* and destroyed in advance all the German attempts to use religious sentiments as disintegrating forces within the USSR. In fact, one of the few liberties that Germany had granted to the populations of the occupied territories was precisely that of returning to some religious practice. Besides, German policy was ambiguous since its aim was the final destruction of religion; at the same time, it tried to use the churches of the USSR as channels of propaganda against the Bolshevik regime. Thus religious practice was revived in the occupied territories and in some cases – notably in the diocese of Pskov – reached a high level of reorganisation. Stalin, for his part, understood the opportunities offered to him by religion. As for the Orthodox Slavs and Caucasians, Orthodoxy was for them a national religion and by allowing it to revive he restored to the nation its historical background. For the other peoples, for the most part Muslims, Islam was an all-absorbing religion, taking in every aspect of their personal and national existence and of their ideas. They had defended it passionately in the 1930s and to restore to them the right to practise it was a solution to a great many of their problems. The *rapprochement* with the Orthodox Church opened the way to change. Institutionally nothing stood in the way of a change of attitude towards the religions. The Constitution of 1936 gave Soviet citizens the right to practise their religion. Of course, in practice, the regime had encouraged the complementary clause of the Constitution, that is to say 'the freedom of anti-religious propaganda'; and the second Five-Year Plan, of the 'League of Godless Militants' (created in 1925 by Yaroslavsky), that had presupposed that by 1937 religion would be completely destroyed in the USSR. However, religion had, even before the war, benefited from the change in the Constitution. In 1935, priests began to return from deportation and to resume their functions. During the years which followed, the bishoprics which were almost all vacant were once again filled. The ordination of priests and the episcopal consecrations grew more numerous, filling the vacancies left by the persecutions. In 1935 Yaroslavsky admitted ruefully that the number of atheists was decreasing and the anti-religious propaganda was of very little interest to the citizens. In 1936 sacrilegious carnivals were forbidden. This movement was interrupted by the great purges which hit the Church hard and led to the closure of almost all places of worship. But by 1941 the situation once again was tending to improve. The annexation of the Baltic States and of a part of Poland and of Bessarabia brought into the USSR millions of Christians – Orthodox, Catholic and Protestant – organised within powerful churches. The difficult rebirth of the churches after 1939 was helped by his contribution of living and dynamic religious forces.

The Orthodox Church itself helped to make the Stalinist reversal easier. As soon as he heard of the invasion the Metropolitan Serge, locum tenens of the patriarchal see since 1925, made his attitude clear in a pastoral letter in which he exhorted all the faithful to defend their country against the invader. Without referring to the government – the head of the Orthodox Church respected the separation of Church and State – he constantly called on his compatriots to fight, introducing into the liturgy special prayers for a threatened country.

The reaction of the regime was equally prompt. In September 1941 the anti-religious periodicals were suppressed, and the 'League of the Godless' dissolved. From Simbirsk where he was evacuated while the German army marched on Moscow the Metropolitan could, after 1942, enter once more into contact with most of the dioceses and send out appeals and concrete proposals to aid the war effort (the offerings of the faithful paid for the armoured column Dimitri Donskoy).

In besieged Leningrad, the Metropolitan Alexis was indefatigable in the life of the civilians and in the defence of the city, showing that there was real participation by the Church in the efforts of the State to save the country.

For its part, the State gradually recognised the place of the Church within the Russian nation. At Easter 1942 the curfew was lifted in Moscow so that the faithful could go to midnight Mass. In the course of the year, the Metropolitans Alexis and Nicholas were invited to take part in the work of commissions set up to inquire into the German war crimes, and they could send out direct appeals which were handed out by the army to the partisans in the occupied territory. The anniversary of the Revolution, in November 1942, was greeted by congratulatory telegrams from the head of the Church, the Metropolitan Serge and from the Metropolitan Nicholas addressed personally to Stalin which were reproduced by the press.

In the name of the clergy and of all the faithful of the Russian Orthodox Church, loyal children of our motherland, cordially and in a spirit of prayer I salute in your person the leader chosen by God of all our military and cultural forces who will lead us to victory over the barbarian invaders, to prosperity in the peace of our country and to the radiant future of its peoples. May God crown with success and glory your great prowess in the defence of the motherland.

The publication of his telegram in *Pravda* on 9 November showed how far the *rapprochement* which took the form of a friendly dialogue between the head of the State and the head of the Church had gone. On 4 September 1943 – here as elsewhere, the turning of the tide at Stalingrad enabled Stalin to follow his policy through without seeming to be yielding to a desperate situation – the three highest dignatories of the Orthodox Church met Stalin in the Kremlin, thus healing the rift which had lasted for so many years. Stalin authorised the election of a new Patriarch to the see left vacant since 1924. The council met three days later and elected the Metropolitan Serge who had in fact been in charge of the Church for the past seventeen years. The Council decided to excommunicate all those who 'had betrayed the faith and the motherland', thus expressing its complete loyalty to the defence of the USSR. This Council was – in spite of its restricted character – of considerable importance in the life of the USSR at war. It confirmed the reconciliation of the State with religion; it gave an incontestable leader to the Church and reintroduced the elective principle into the social non-political life of the USSR. To be sure, the election of the Patriarch was carried out through acclamation and not in the forms envisaged by the Council of 1917; nevertheless it was an election and not an appointment, and this in an hierarchy independent of the State and of the Party and representative of a creed alien to the Party. For the first time for years the ideological monolithism seemed to retreat.

The Patriarch died some months later on 5 May 1944, but he left behind him a restored Church, with its own legal newspaper – *Review of the Moscow Patriarchy* – and an accepted status, since a new council met in January 1945 to decide on his successor. Earlier the president of the Council in charge of the affairs of the Orthodox Church, Karpov, had declared to the religious authorities:

What is taking place in the relations between the Church and the State is neither fortuitous nor temporary, nor is it a tactical manoeuvre ... the measures taken by the Soviet government of the USSR conform perfectly to the Constitution of the USSR and are a mark of approbation for the attitude taken by the Church towards the State in the ten years which preceded the war and above all during the war.

These measures were a veritable oral Concordat which allowed the Church to organise itself and to root itself once again in the USSR, and which enabled the government to appear as the real unifier of the USSR internally, and externally, above all among the Slavs, to assume a unifying aspect by strengthening the elements of ethnic and cultural solidarity. The *rapprochement* with the Russian Orthodox Church was accompanied by an attempt to appease the Muslims. The policy of national *détente* led, in the Muslim republics, to a gradual renaissance of religious life which was crowned in October 1943 by the creation of a 'Central Direction of Muslims' in Tashkent. The installation of the Mufti of Tashkent which was a sign to the faithful of the goodwill of the Soviet authorities towards Islam as an organised religion not only had beneficial effects among the nations of the USSR but damaged German plans in the Caucasus. The German authorities tried to weaken the Soviet position by setting up instead of the Mufti of Tashkent a mufti of their own creation; but they were unable to find a credible candidate and their agitation helped to demoralise those Muslims who until then had been tempted to rely on Germany. The Jewish community itself benefited from this change, again above all as a national and not as a religious community. The instrument of the co-operation between the authorities and the community was the Jewish Anti-Fascist Committee, set up as a bridge between it and the Jews abroad, for the most part American. Here, the attitude towards the Jews was above all inspired by considerations of foreign policy.

The condition of the Jewish community of the USSR at the moment of the invasion was very critical. A large part of the community had been taken by surprise by the German advance into the Ukraine and Byelorussia, where it lived and where countless Jews were exterminated. Those of the Crimea, more fortunate, were sometimes evacuated in time, with their whole kolkhozes, and dispersed throughout the non-occupied territories. Nowhere in the USSR were there proper reception centres for the Jews because the life of their community had been reduced to a minimum in the pre-war years. Since 1938, the Jews had had no proper organisation, publications nor schools. However, the invasion was to revive and reassemble the Jewish community who as a social body in Soviet life reappeared on 24 August 1941 at a 'public meeting of the representatives of the Jewish people' which took place in Moscow. Ehrenburg's declaration showed clearly that this meeting was indeed a resurrection of the community. 'I grew up,' he said, 'in a Russian town. Like all Russians I defend my motherland. But the Nazis have reminded me of something else. My mother was called Khana. I am Jewish. I say this with pride. We are those whom Hitler most hates and that is an honour for us.' In taking note of the meeting and in reporting Ehrenburg's speech, *Pravda* (25 August) showed that the government accepted the revival of genuinely Jewish cultural life. From the beginning of September 1941, two leaders of the Polish *Bund* – Henrich Erlich and Victor Alter – who had just been freed from their Moscow prison, relying on the incipient liberalism, proposed to Stalin and Beriya the creation of a Jewish committee. The promoters of the project encouraged by Beriya wanted to create an instrument of solidarity and of mutual help for all the Jews in occupied Europe and wanted at the same time to organise the life of the dispersed Jewish community in the USSR. The project which its authors thought had been approved was buried and Erlich and Alter were arrested on 5 December 1941 and executed shortly afterwards. In April 1942, however, a Committee of Jewish Intellectuals of the USSR was created which set up a Jewish Anti-Fascist Committee and organised a second 'public meeting of the representatives of the Jewish community' on 24 May 1942. The Committee set up in the spring of 1942 was very unlike the one of which Erlich and Alter had dreamt a few months earlier. The Committee's programme was uniquely concentrated on the war effort and the help which the Jews could give to the USSR. The Committee collected money, appealed to the foreign Jewish communities and set itself the immediate goal of buying for the Red Army 100,000 tanks, 500 aeroplanes and a warship. In 1944 the Committee was able to hand over to the Red Army nearly 3 million dollars collected in the United States. With a solid organisation, a newspaper, *Eynikayt*, assembling the élite of the Jewish intellectuals and artists of the USSR, the Committee during the war could act as a genuine organ of foreign propaganda convincing especially the large Jewish-American community of the change in the USSR. As for the fate of the Jews in the USSR, the Committee, in spite of various attempts by its members to persuade the government to pay more attention to their situation, systematically ignored it. Thus the Jewish Anti-Fascist Committee set up in 1942 had less freedom of action as Jews than the purely religious organisations (Christians and Muslims), which profited from the easing of the government's attitude towards them genuinely to renew an active religious life, to restore the places of worship, to reconstitute the clergy and set up a stable structure. The 5 million or so Jews, on the contrary, were unable to rebuild either an educational structure or information organs; but the existence of the Committee and its newspaper and the regrouping of the Jewish intelligentsia gave reasons to hope that the ground had been prepared for a post-war cultural renaissance.

The government's attitude towards the religious and cultural communities was thus more subtle than has generally been supposed. As for the nationalities, the Soviet government in its religious policy during the war tried to respond to the immediate problems and at the same time to keep the post-war perspective in mind. Its concessions seen overall are less important than they appear if taken singly. It is easy to understand the favour shown to the Orthodox Church between 1941 and 1945, if one remembers that historically it was inseparable from the Russian nation which affirmed itself and was developed in the fight against the non-Christian elements first and non-Orthodox later. Orthodoxy is an integral part of the history of Russia; its renaissance could only help the effort to save the nation during the war and to prepare, when the post-war decisions were being worked out, the unification of Eastern Europe in part Orthodox under Soviet authority. Here again post-war concerns played a fundamental role. By contrast, the non-assimilated Jewish community whose foreign links and personal awareness had been increased by the Nazi persecution, for whom a national State was also a possibility, only found a limited place in this search for the elements of Russian and Soviet national coherence, which characterised Stalin's war policy.

But the decisive element in the war effort remained – Stalin as well as the Germans understood this – the Soviet peasants, who held in their hands the material survival of the urban population and of the Red Army. During the Civil War, the peasants had remained aloof from the conflict; in the war of 1941–45, their active contribution was indispensable and Stalinist policy made an unprecedented effort to conciliate the peasants which, in conjunction with German mistakes, was to draw them over to the side of the Soviet regime.

As well as the loss of the agricultural provinces in 1941, other elements made for weakness in the countryside: the mobilisation in the countryside, the lack of organisation of transport, the requisitionings of the army and the inflation caused by the cost of the war. In 1940, the government had begun to develop food-producing crops in the east but these measures which showed some concern about the future of German-Soviet relations were still inadequate. The problems of the labour force were crucial. For comparable areas, the countryside in the unoccupied regions in 1942 had at its disposal only 68.6 per cent of the workforce of 1940, and women who then represented 52 per cent of its workforce, made up 71 per cent at that time. The army requisitioned horses and tractors. The yields fell from 700 kg of cereals to the hectare in 1940 to 420 kg between 1943 and 1945. The fall in the harvest, the losses suffered in livestock were all the more serious in that the needs of the army and the towns were increased by an influx of refugees from the occupied territories or mobilised from the countryside. The levies from the countryside had thus to increase at the very moment when the political supervision of the peasantry was diminished because of the departure for the army of the local Party members. The co-operation of the peasants had thus to be won voluntarily, no matter what the cost, and owing to lack of manpower and for psychological reasons the government could no longer use constraint. The government's policy was to ask the peasantry for an increased effort within the collective structure and in 1942 to give it at the same time a greater freedom to sell its personal produce. The government increased the number of work days of the peasant by about 50 per cent (254 in 1940, 352 in 1942) which in principle left the peasants less time to look after their own plots. The State tried at the same time to stabilise agricultural prices and in 1941 and 1943 and increased the agricultural tax. But it made up for these measures by turning a blind eye to the growth of the peasant market. This grew rapidly for the demand in the towns in which rationing had been imposed since October 1941 was considerable. The peasants, in spite of their lack of time, increased their personal crops to their maximum. In some regions, in Central Asia particularly, the crops destined for the market spread. The kolkhoz market increased its activity. In 1944 it supplied half of the food purchase as against 25 per cent in 1940 and the gap between its prices and those paid by the State was considerable (the differences registered in Moscow in 1943 ran from 1 to 30). Taking advantage of of the government's tolerance, the peasant thus increased his efforts and this enabled the food supplies of the towns to be maintained at a relatively stable level.

The policy towards the peasants during the war had many sociological consequences. First of all, the war reconciled the town and the countryside and to some extent rekindled in the inhabitants of the towns the memory of their rural origins. The inhabitants of the towns were not content to queue in the markets to which the peasant sometimes found it difficult to deliver his produce because of transport difficulties, but went themselves into the country. Family links were often reforged in this way. Businesses also organised provisioning services by linking themselves with the peasant crops. Lastly, the town dwellers, whenever possible, cultivated gardens which sometimes enabled them to appear in the market as producers. The agricultural crops of the urban population rose from 5 million in 1942 to 11.8 million in 1943 and to 16.5 million in 1945. Two-fifths of the families cultivated their gardens and harvested from 400 to 600 kg of potatoes and vegetables.

As well as this reconciliation between town and country, there were two important psychological changes: the weakening of the collective conscience and the weakening of the value of money. The rise in prices meant that in the war years the urban population faced a growing gap between prices and wages. While the nominal wage increased by 29 per cent between 1940 and 1945, the retail price index was fixed at 325 in comparison with pre-war. The situation of the urban population was very difficult but the figures did not always represent reality because the citizens had to find supplementary means of existing. In the country, on the contrary, a relatively privileged situation developed because the peasants, in spite of the decrease in the payments in kind (110–180 kg of cereals and 70–190 kg of potatoes in 1940), drew from the market infinitely higher sums than in the past. But everywhere, because of inflation, money had lost its value and the old practice of payment in kind and above all of barter had gained ground. In the sovkhozes and the MTS bonuses in kind were introduced and in the markets barter flourished. In some markets 6 kg of flour could buy a silver mug.

The peasantry were to be profoundly marked in the future by the choices made by the government. In fact, the freeing of production prices during the war brought about not only a shift towards the private sector but also a devaluation in the work in the kolkhozes. The prices given by the State during this period were completely unrelated to the real costs of production. A consensus on prices was established because the members of the kolkhozes compensated for them by developing the free sector and saw the government's tolerance as an implicit recognition of the situation. But at the same time, the idea that the kolkhoz would only play a minimal role in peasant life (in 1945 90% of the peasant income came from the free market) developed and fostered an attitude of indifference towards the kolkhoz. The consequences of this psychological attitude was to be the running down of the goods and equipment of the kolkhozes and the low productivity of the collective work, problems which were to confront the authorities when the time to rebuild the Soviet economy arrived. But in the war years, what was essential was the loyalty of the peasant to the common effort and this loyalty was obtained by the government by staking everything on the private interests of the peasant.

The Stalinist action during the war was not only conciliatory towards the most vulnerable social groups, it also had strictly political aspects concerning the structure and the form of government. This evolution of the government's attitude took two forms. On the one hand, at least at the military level, there was a division of powers and the introduction of a greater rationality. At the same time, power became more and more personalised and the personage of Stalin, absolute master of the USSR, definitely took shape at that time.

After the invasion Stalin tried to give the economic and military authorities some initiative by freeing them from political controls. At the request of the managers, the political authorities were restricted to a purely ideological role, and in some cases eliminated altogether from factory sites. The organisational problems of the movements of personnel, were henceforth in the hands of the technical authority. Even the political speeches were often reduced or suppressed because they encroached on the work. Thus in the war years there was, in the name of efficiency, a decline in the role of, and the opportunities for intervention of, the Party in the economy of the USSR. The same was true at the military level at which not only the political commissars disappeared but the political authority took second place to the army leaders. Khrushchev stated that Stalin's strategy was on a global scale, but all the statements of the Soviet army leaders contradict this statement. Stalin, on the contrary, left it to the generals to draw up their battle plans on the front, said Marshal Konev, and the intervention of his GHQ in the decisions of the High Command seemed to have been limited to suggestions and meetings. To be sure, Stalin promoted himself to the rank of Marshal, then in 1945 of Generalissimo of the USSR, but after the initial defeats he gave the army a fairly free hand to develop its own ideas and spared them the political controls of the past.

At the highest level, bodies were set up to rationalise the action of the State, free from any interference from the Party. The State Defence Committee, set up a few days after the invasion, was to co-ordinate the whole edifice, the ministries, the army and even the party. The creation of this unique body showed a determination to give priority to the technical aspect of the problems, to ease the political controls which made the functioning of the various bodies very heavy and complex. The economic organisation was, in the same way, placed entirely in the hands of an economist, Voznesensky, who was given the task of solving all the problems caused by the demands of the war.

Parallel with this extension of authority, or rather this depoliticisation of the immediate action, another change took place at the highest level, the growing personalisation of Stalinist power. Stalin in the war years created his future image for his people and for the external world.

The personalisation sprang first from an official concentration of powers in Stalin's hands. On 6 May Stalin had replaced Molotov as President of the Council of Peoples' Commissars, gathering into his hands for the first time since 1917 the powers of the State and the Party, traditionally dissociated. To be sure, this was only the official recognition of a situation which had long existed in practice; nevertheless, it was an important change in that Stalin was to apppear increasingly as the head of State, thus accentuating the temporary eclipse of the Party. To the world outside Russia, too, this attitude was reassuring. Stalin appeared more than before as a traditional head of State and he seemed to have forgotten the universal mission of the Party.

The Stalinist personage took shape at the time. Bukharin's formidable opponent, the retiring man of the congresses, transformed himself into the protective hero. The 'Little Father' of the people became in part the traditional image of the protective authority. This evolution, which was taking shape before the war, was fostered by the conditions of the war. In October, when the German army advanced on Moscow at the same time as it was threatening Leningrad in the north and Kiev in the south, the regime seemed to be doomed. In Moscow, the authorities were panic-stricken, the NKVD burnt its files, those who held any official position fled, looters ransacked the shops. Confronted with the general panic, Stalin's reaction showed no signs of his previous collapse. He proclaimed a state of siege, organised the evacuation of the government, the Party and the diplomatic corps to Kuibyshev, sent Zhdanov and Voroshilov to Leningrad and Khrushchev to the southern front. In the deserted Kremlin, he remained alone with the GKO and the depleted Politburo, as some of its members had been sent to other fronts. For the Muscovites during the whole period which brought the Germans to within 20 km of the capital, Stalin's presence was the symbol of resistance. His image obliterated that of the government which had been evacuated to a safe place, but also of his colleagues in the Politburo who had remained in Moscow but whose personal role was too vague to make any impression on the population. Although during the months of the retreat of the Red Army the press was reticent about Stalin's military role, from the first German reverses the Father of the people insisted on his role as war leader, and slowly the revolutionary in his cloth cap was replaced in the official portraits by the Generalissimo. Physical changes also helped to impose the new Stalin. As his hair turned grey, Stalin lost his youthful harshness, his commonplace features were ennobled, and this change lent plausibility to the picture of the caring Father, concerned for the common fate.

Humanised, his personage also assumed a new dimension; it embodied the national pride to which appeal was constantly made. He embodied it as a result of the victories won by the Red Army; he embodied it by the place he assumed in world affairs. Relations with the Allies were in Stalin's hands and for years he fought a hard fight with them on the opening of a second front. Unable until 1943 to respond to this demand, the Allies showered him with praise. 'The hopes of civilisation rest on the banners of the Red Army', said MacArthur; such compliments strengthened his moral position. In 1943, Stalin decided with Churchill and Roosevelt the future fate of Europe. The humiliation of Munich where everything had been decided without the USSR was wiped out. Now Stalin imposed his will, made the Allies come to the frontiers of the USSR because he would not agree to travel himself. His demands were, for the Soviet public, the revenge of a young State ignored for years by the world outside and whose efforts were suddenly of vital importance to that society. They were also the justification for Stalin's policy and the public was ready to forget his excesses and his mistakes. His allies also accepted this picture of the genial leader and concluded that 'Uncle Joe was a good chap'.

The evolution of the Stalinist personage which obliterated the memories of the period of terror corresponded with the general evolution of the USSR. The German invasion had united the people around the regime. Stalin was able to symbolise this unity and identify himself with it, and because he was in power he had the advantage of being seen in a favourable light by most of the population. The character of arbiter which he had maintained for so long had given his real role in the establishment of the system of terror a certain ambiguity. Above all, in spite of his natural brutality, he had always given the impression of a leader motivated solely by a passion for politics and devoid of personal passions and interests. This impression of a modest man who identified himself with his country and reflected it also helped to rally the Soviet people around him.

The four years between the German invasion and the rapid advance of the Red Army across Europe towards Berlin were for the USSR years of incredible suffering, of extraordinary efforts and of profound changes. In 1941, the Soviet regime had seemed to be doomed and its leaders had feared that there would be a general collapse. In 1945, the USSR which emerged from the war was a powerful State internationally if not economically, and for the first time a united nation. Until 1940, the Soviet State was the organisation of diverse peoples united by the same system, but whose unity was anything but apparent; it was also a conglomeration of social groups broken by the structural changes and the terror from which each element, withdrawn into itself, sought above all to protect itself. The 'Great Patriotic War', rightly named in that it had created a Soviet motherland, had forged its peoples and the groups into a community. Of course, the Soviet community which emerged did not solve all the problems and the growth of nationalisms and, the development of non-political hierarchies were to create difficulties after 1945. However, these future difficulties could not obscure the fact that the Soviet people had acquired between 1941 and 1945 a new unity around an idea which was new to them, that of the motherland.

CHRONOLOGY OF THE GREAT PATRIOTIC WAR

	WAR IN THE EAST	WORLD SITUATION
1939		
23 Aug.	Nazi-Soviet Pact.	
1 Sept.		Germans invade Poland. France and Britain declare war on Germany.
30 Nov.	USSR invades Finland.	

1940

12 Mar.	Russo-Finnish Peace Treaty.	
9 Apr.		**Germans invade Denmark and Norway.**
10 May		**Invasion of Holland, Belgium and Luxemburg.**
10 June		**Italy enters the war on the German side.**
22 June		**Franco-German Armistice.**
17–23 June		**USSR invades the Baltic countries, Bessarabia, northern Bukovina.**
July, Aug. Sept.		**Battle of Britain.**

1941

6 May	Stalin, head of government.
22 June	German invasion.
28 June	Germans occupy Minsk and a part of Lithuania, Latvia and the western Ukraine.
3 July	Stalin's first speech after the invasion.
12 July	Signature of the Russo-British pact of mutual assistance.
14 July	German advance on Leningrad.
16 July	Fall of Smolensk, beginning of the German advance on Moscow.
17 Aug.	Occupation of Dniepropetrovsk.

Stalin

1941

8 Sept.	Beginning of the siege of Leningrad.
30 Sept.	Beginning of the battle of Moscow.
2 Oct.	Fall of Orel.
12 Oct.	Fall of Kaluga.
13 Oct.	Fall of Kalinin.
16 Oct.	Fall of Odessa.
20 Oct.	State of siege proclaimed in Moscow.
24 Oct.	Fall of Kharkov.
25 Oct.	German offensive halted before Moscow.
30 Oct.	Beginning of the siege of Sebastopol (nine months).
3 Nov.	Fall of Kursk.
16 Nov.	Beginning of the second offensive against Moscow.
19 Nov.	Fall of Rostov.
29 Nov.	Liberation of Rostov.
5 Dec.	Eden arrives in Moscow.
6 Dec.	Beginning of the Russian counter-offensive at Moscow.
7 Dec.	Japanese attack on Pearl Harbor.
8 Dec.	The United States and Britain declare war on Japan.
11 Dec.	Germany declares war on the United States.
25–30 Dec.	Liberation of Kaluga.

1942

10 Jan.		Japanese invasion of Dutch East Indies.
8 May	German attack in the Crimea.	
20 May	Fall of Kerch.	
26 May	Molotov goes to London and Washington.	
3 July	Fall of Sebastopol.	
28 July	Germans retake Rostov.	
12–15 Aug.	Churchill–Harriman–Stalin Conference in Moscow.	
13 Sept.	German attack on Stalingrad.	
23 Oct.		Beginning of Battle of El Alamein.
8 Nov.		Allied landing in North Africa.
22 Nov.	300,000 Germans encircled at Stalingrad.	
21 Dec.		Eighth Army reaches Benghazi.

1943

2 Jan.	German evacuation of the Caucasus begins.	
23 Jan.		Eighth Army reaches Tripoli.
26 Jan.	Liberation of Voronesti.	
31 Jan.	Surrender of Field-Marshal von Paulus at Stalingrad.	
8 Feb.	Liberation of Kursk.	
14 Feb.	Liberation of Rostov.	
16 Feb.	Liberation of Kharkov.	
15 Mar.	Germans retake Kharkov.	

1943

7 May		Allies take Tunis and Bizerta.
12 May		Surrender of German army at Tunis.
22 May		Dissolution of the Comintern.
10 July		Allied landing in Sicily.
12–15 July	Russian counter-offensive at Orel.	
26 July		Fall of Mussolini
5 Aug.	Liberation of Orel and Bielgorod.	
16 Aug.		Americans take Messina.
8 Sept.	The Russians recapture the Donbass.	
10 Sept.	Liberation of Mariupol.	
25 Sept.	Liberation of Smolensk.	
13 Oct.		Italy declares war on Germany.
18 Oct.	Foreign Ministers' Conference in Moscow.	
6 Nov.	Liberation of Kiev.	
12 Nov.	Liberation of Zhitomir.	
19 Nov.	German counter-offensive and recapture of Zhitomir	
28 Nov.	Beginning of the Teheran Conference.	

1944

27 Jan.	Final liberation of Leningrad.	
4 Mar.	Soviet offensive in the Ukraine.	

1944

2 Apr.	Soviet troops enter Romania.	
11 Apr.	Beginning of the liberation of the Crimea.	
9 May	Liberation of Sebastopol.	
13 May	Whole of Crimea liberated.	
4 June		Fifth Army enters Rome.
6 June		Normandy landing.
10 June	Soviet offensive in Finland.	
3 July	Liberation of Minsk.	
18 July	Rokossovsky's army enters Poland.	
20 July		Attempted assassination of Hitler.
1 Aug.		Relief of Warsaw.
15 Aug.		Allied land in the South of France.
25 Aug.		Liberation of Paris.
9 Sept.	Russian troops enter Bulgaria.	
12 Sept.	Russo-Romanian armistice.	
29 Sept.	Russian troops enter Yugoslavia.	
20 Oct.	Liberation of Belgrade.	
2 Dec.		De Gaulle arrives in Moscow.
16 Dec.		German offensive in the Ardennes.
27 Dec.	Soviet troops encircle Budapest.	

Stalin

1945

12 Jan.	Beginning of the great Soviet of offensive in Poland.	
17 Jan.	Fall of Warsaw.	
19 Jan.	Fall of Cracow.	
23 Jan.	Russian troops reach the Oder.	
4 Feb.		Opening of the Yalta Conference.
13 Feb.	Budapest falls.	
29 Mar.	Soviet troops cross the Austrian frontier.	
12 Apr.		Death of President Roosevelt.
13 April	Soviet troops take Vienna.	
16 Apr.	Opening of Soviet offensive on Berlin.	
27 Apr.		Meeting of American and Russian troops at Torgau.
30 Apr.		Hitler commits suicide.
1 May		Surrender of German army on the Italian front.
2 May	Fall of Berlin.	
7 May		Unconditional surrender signed by Jodl at Eisenhower's HQ at Reims.

1945

8 May	Unconditional surrender signed by Keitel at Zhukov's HQ near Berlin.	

9 May	Fall of Prague.	
	VE Day in the USSR.	
17 May		Opening of Potsdam Conference.
8 Aug.	USSR declares war on Japan.	
2 Sept.		Japan capitulates.

THE RECONSTRUCTION OF THE ECONOMY

When the war ended in 1945, in spite of the huge Soviet losses, new elements in the internal and external situation of the USSR enabled the Russians to hope that the political system would relax.

The war had forced the government to relax its authority and to diminish the constraints which weighed upon the citizens. The national and religious renaissance of the years 1941–45, and the intellectual liberation had profoundly changed the internal climate of the USSR. Moreover, the struggle and the victory had brought together the leaders and the masses. The effort had been a common one, it had forged a Soviet nation around those who had led it and never, since the Revolution, had the government enjoyed such popular support which had swept away the bitterness and the scepticism of the pre-war years.

At the same time, the international situation of the USSR changed. Its expansion in 1939, the erection of the East European glacis, the rapid extension of the communist system first of all in Yugoslavia then in the other countries liberated or conquered by the Red Army, put an end to the State of the 'besieged fortress'. The USSR had friendly neighbours; she occupied a leading place on the international stage, the external class war which justified the maintenance of the authoritarian system at home seemed to have weakened. Could the *détente* of the war years become the basis for new political relations in the USSR?

However, these favourable elements were outweighed by negative factors, all linked to the war itself: the human losses had been very great, the economy was in difficulties, and the Soviet system was under stress. To some extent the simplest and most essential task was that of reconstructing the economy, in order to hearten a people which had paid very dearly for the war.

THE WAR AND SOVIET SOCIETY

The human cost of the war can only be established approximately by comparing statistics, since the government has refused to publish the figures. It is generally estimated that there were 18 million dead, about 7 million in the armed forces and 11 million civilians, 8 million of whom were in the occupied zone. These estimates take into account the territorial acquisitions of the USSR after 1939 which means that its population in 1945 was, in spite of everything, higher than before the war. In this terrible balance-sheet no account is taken of the physical sufferings undergone by a part of the population and which diminished it for ever; the wounded, and famished of Leningrad, etc. The children especially who lived through the siege of Leningrad were definitely marked by the ordeal. The years immediately after the war prolonged the sufferings of the Soviet people, especially in the cities. For a long time, it was difficult to guarantee the country's food supplies. After 1943, the retreating German armies carried out a scorched-earth policy, destroying the crops, the livestock, and burning the villages. The harvest of 1945 was a poor one, that of 1946 was, because of the serious drought, catastrophic, below the worst pre-war harvests. Famine returned to the USSR immediately after the war and rationing had to be maintained until 1947. Owing to the shortages, prices began to rise and the government tried to adjust the discrepancy between the official trade and the free market. In September 1946 the cost of the officially rationed food rose by 90 per cent (50% for meat, 365% for flour). The government tried to compensate for the rise by increasing earnings below 300 roubles a month by 36 per cent, those which were above 700 roubles by 10 per cent, so that the lowest categories were helped; but this increase was not enough to keep up with the rise in prices. In December 1947, the government decided to reduce inflation by a currency reform, the first since 1924. The notes in circulation were exchanged at the rate of 10 old roubles for one new one, savings accounts up to 3,000 roubles were converted at their nominal value and at the rate of 3 old roubles for 2 new ones for the bracket between 3,000 and 10,000 roubles; the current accounts of the kolkhozes were converted on the basis of 5 old roubles for 4 new ones, and loans at 3 roubles for 1.

The intention of the reform was to distribute the purchasing power between the town and the countryside and to curb the 'speculators', especially the members of the kolkhozes who were suspected of having made too great profits out of the shortages. The mass of notes in circulation which were not presented for exchange is set at 30 per cent, as many feared that they would be questioned about the origin of the notes. After 1948, a fall in many retail prices coinciding with the end of rationing improved the conditions of the people, increasing their purchasing power and their right to purchase. Moreover, the depletion brought about by the currency reform in the peasants' savings brought a fall in the prices on the free market which, in 1948 and 1949, fell to the same level as the prices on the official market, and in 1949 potatoes could be bought in the Moscow market more cheaply than in the State shops.

Although the food situation improved after 1947, housing in the cities remained very difficult. Compared with the pre-war situation, the problem was worse because of the destruction of whole cities (in 1945 nearly 20 million people had lost the roof over their heads), and also because so many peasants had migrated to the cities, a movement which accelerated after 1946. A movement of population also took place towards regions which the government was trying to develop and in which there were better-paid jobs: Siberia, and the Urals. Cities grew up very rapidly, Novosibirsk, where the population quadrupled in a few years, Omsk, where it doubled, Chelyabinsk, Magnitigorsk, Sverdlovsk. But the difficulties in settling in the huge inhospitable cities, in which housing and food supplies were as uncertain as everywhere else, made for great instability. Every four or five years, the pioneers returned home, giving way to new arrivals. One of the most difficult problems for the government was to settle the population in the eastern regions.

THE RECONSTRUCTION OF INDUSTRY

During the war, Soviet industry faced great difficulties, but in 1945 its foundations remained solid. The government had protected it; as the Germans advanced, the government had evacuated cities, dismantled factories and set them up again in the liberated territories as the occupying troops retreated. In 1942, production began again in the Donbass and the steppes, and when the moment of liberation came every factory in an unoccupied city took charge of an occupied factory and made itself responsible for starting it up again by supplying it with equipment and manpower. However, in 1945, it was mainly in the equipment which had been used until it was worn out, that the after-effects of the war took their toll. The production of coal, electricity and steel, after a precipitous fall in

1942, picked up rapidly in 1943. The production of oil recovered more slowly.

As regards people, conditions favourable to recovery existed. The working class had not failed the regime during the war, and had proved its attachment to it. The townspeople had been mobilised and had been led by circumstances to use their own initiative; yet the war had not changed relations within the factories, although a greater degree of co-operation between the working class and its leaders had been established. The problem of industrial reconstruction was thus essentially technical: it was primarily a problem of manpower.

This is probably the explanation for the rapidity of the industrial reconstruction after 1945. In 1947, in spite of an immense effort, industrial production was still below its pre-war level; but in 1948 the level of 1940 was surpassed everywhere and was doubled in 1952. This spectacular recovery did, however, give rise to some problems, mainly in heavy industry. As in the past, consumer goods had been neglected. But at the end of the war, the people were urgently in need of various necessities. In 1948, it was impossible to buy a pair of shoes in the USSR. It was not until 1952 that the production of consumer goods reached the pre-war level again – a level which itself had always been notoriously inadequate and which after the war was strikingly so, because everything had been used up and never replaced during the years of conflict. In spite of the overall recovery of production, the material poverty of the citizens was greater than it had ever been. It is estimated that in 1948 the real wages of the worker or of the average employee was at the index 45 in relation to the index 100 in 1928; in 1952 it reached the index 70.

This led to another problem, the gradual erosion of the solidarity born of the war between the governors and the governed. In 1945, the Soviet people hoped for a widespread political and material change. Many of the Soviet people had become aware, because of the war, of the gap which existed between living conditions in the USSR and those abroad, and in spite of the government's precautions to isolate those who had been demobilised and the repatriated deportees, what they had seen outside the USSR was soon known to everyone. Victorious at the cost of immense suffering, the Russians wanted to reap some benefits from their sufferings. They agreed in 1945 to make yet another effort so that the national economy would recover. But by 1947 the people in the cities had again lost heart. The stress laid on the reconstruction of heavy industry alone, the continuing difficulties in daily life all obscured the effects of victory. The political climate in the Russian cities deteriorated again at this time and this brought an increase in the political control of the working class, which had been relaxed during the war. Once again, the USSR entered the vicious circle of pressures and grievances and the *détente* of the war years became blurred.

THE REORGANISATION OF AGRICULTURE

In spite of inequalities and black spots, the industrial reconstruction was one of the post-war achievements. But this could not be said of the agricultural sector which once again was at the heart of the difficulties of the USSR. Here, the problem was not simply a technical one, nor one of rebuilding on existing foundations. The task, which had been carried out with such difficulty since 1929, of building a socialist agriculture had been seriously compromised by the war, and the government had to contend with technical problems but above all with the eternal adversary, the peasant. Clearly agricultural production had to be set to work again but first of all the peasant had to be 'brought back to the fold', after several years of comparative freedom which had encouraged his individualistic tendencies. Moreover, unlike the situation which existed in industry, it was not a successful pre-war situation which was recreated but a forcible integration which had never been completely accepted and which the war had shown to be extremely precarious.

The problems were of various kinds. The first was that of the fall in production during the war years. The soil had been ravaged by the fighting, by being left untended and by the policy of systematic destruction of the German authorities. The agricultural equipment was reduced and what remained had been considerably damaged. Lastly, the livestock had been greatly depleted. At the end of the war, the USSR had barely 10 million horses and as many pigs compared with the 20 and 28 million in 1940. The stocks of cattle and sheep at their lowest in 1943 began to increase but there was a shortfall of nearly 10 million cattle (15% of the herd) and 25 million sheep (nearly 25% of the total). This problem was all the more serious in that the territorial annexations had slightly increased the population of the USSR. In the sphere of production, progress was extremely slow since for a considerably enlarged territory, the production of cereals in 1950 reached a level which barely equalled a normal pre-war year and the livestock remained at the same time between 16 and 18 per cent below its pre-war level, while the population of the USSR continued to grow (1% per annum between 1945 and 1953, that is nearly 2 million per annum).

The problem of agricultural production was affected not only by the destruction but by the evolution of the structures of agriculture. The decollectivisation brought about by the Germans or spontaneously by the peasants had resulted in an often irrational breaking up of the land and the dispersal of the equipment of the kolkhozes. The growth of individual plots recovered from the collective land had brought a change in production and produce was drained off towards the free market at the expense of the official market, thus once again creating a problem of prices for the consumer. The need to rebuild the collective capital led to clashes with the peasantry which did little to improve its goodwill and thus to encourage it to produce more; on the contrary, the political pressures put up the back of the peasant and caused him to concentrate on his own concerns and his plot. In the countryside also, the reconstruction put an end to *détente*.

Lastly, the government had to deal with a complex problem of the political education and supervision of the peasantry. The period of the war had been marked by the growth of 'kolkhoz administration'. This term in fact concealed a proliferation of parasites who had lived out the war on the kolkhozes, black marketeers, bureaucrats, etc. Inquiries made by the Soviet authorities after the war revealed the existence in the kolkhozes of numerous individuals often completely unknown, but whom the peasants had tolerated because they had guaranteed them a real measure of independence. The problem of the technical training of the peasant had to be completely revised, especially as the rural organisations of the Party were extremely weak and did not guarantee efficient political supervision. A threefold reconstruction thus had to be undertaken: reconstitution of the property of the kolkhoz, setting up of communist organisations in the kolkhozes, and the development of production and peasant education and supervision. The reconstitution of the property of the kolkhozes seemed to be an indispensable preliminary to the recovery of agriculture. The problem was complicated by the need to integrate into the structures of the kolkhozes the population of the conquered regions.

A common resolution of the Council of Ministers and of the Central Committee of the Party of 19 September 1946 created the body which was to undertake the reconstruction. The Special Commission for Kolkhoz Affairs, was presided over by A. Andreyev who was very experienced in agricultural matters. In the past, he had brought to a successful conclusion the collectivisation in the Caucasus and from 1943–46 he had been Commissar for Agriculture. One of Stalin's close collaborators whose confidence he shared with Malenkov, Andreyev, in 1946, was given considerable means to carry out the task entrusted to him. A body of civil servants, some from the Party, others agricultural technicians, were to be sent to the countryside, brought into contact with the local organisations of the Party and given the task of patiently reassembling what had been dismantled. At the end of 1947, 4.7 million ha had been restored to the collective economy. However, the policy of rebuilding the capital of the kolkhozes was gradual; it tried to take account of local problems and of the difficulties in production. The authorities were concerned not to clash with the peasant; this was why in the Baltic countries the individual farmer enjoyed a degree of tolerance until 1949. In Latvia, in 1948 only 2.4 per cent of the cultivated land was part of the kolkhozes; in 1949, this had risen to 76.7 per cent and by 1950 to 90.7 per cent. In Estonia, the collective farms rose from 5.8 per cent at the beginning of 1949 to 75.9 per cent in 1950 and 92 per cent in 1951. In the same way, up to 1950, the government encouraged the rebuilding of privately owned livestock which was seen as the best way to ensure that the USSR had adequate stocks, and in fact during the five years immediately after the war progress in this field was very satisfactory. The individual plots, after they had been put in order, also grew up to 1950 by about 1 million ha. The government could not deprive the people of privately owned production, while the food supplies were still so precarious.

This relative tolerance towards the peasants was however outweighed by the currency measures taken against them in 1947. The savings accumulated thanks to the free market were wiped out by the inroads of 1947 and the fall in official prices at the same time favoured the town dwellers rather than those who lived in the country. After 1948, because of this policy the standard of living in the towns rose more than that in the country, and compared with 1928 (index 100) the real revenues of the peasants stood at 50 in 1949 and 60 in 1952. The relation between town and country was reversed. The peasant discontent made the problem of its supervision and training even more pressing.

The setting up of communist organisation in the kolkhozes was necessary because of the situation existing at the end of the war (only one kolkhoz out of eight had a Party organisation), and because of the pressures which were looming over the life of the peasants. By 1953 communist organisations existed in five out of six kolkhozes. This operation had been carried out in three stages. From 1945 to 1946 a great many communists returned to their villages, demobilised soldiers who had joined the Party when they were in the army. An examination of the state of the rural organisation in the Orel region, as presented in Table 2, is illuminating.

Table 2.   Rural communist organisations in the Orel region

Types of organisations	Numbers of organisations		Numbers of members	
	January 1945	*April* 1946	*January* 1945	*April* 1946
Kolkhozes	34	253	298	1,582
MTS	73	89	557	996
Village	259	536	2,072	7,539
Total	366	878	2,927	10,117

This table which reflects the widespread situation in the USSR in the years 1945–46 shows the growth of the communist organisations and membership in the country in the months after the war; it also reveals the persistent weakness of the training in the kolkhozes as compared with that of the villages. In 1947, in spite of the progress made, less than one-quarter of the kolkhozes had Party organisations. In 1946, the Communist Party directed all its energies towards this problem and tried to move the cells of the village cadres to the cadres of the kolkhozes, to move the members with administrative responsibilities at the village level to positions in the collective farms, and finally it moved them from one farm to another; the final objective was that every farm should have a communist cell.

The second phase of the training of the kolkhozes was to last from 1947 to 1949; it was marked by the speeding up of the redistribution of the existing communist cadres in the countryside. On the one hand, a real increase in the number of cells in the kolkhozes could be seen, and on the other it was reckoned that the number of communists in the countryside had hardly changed.

Thus, in the Ukraine the number of kolkhoz cells quadrupled as compared with the pre-war figure but the membership did not increase. The effects of this policy caused the Party some concern. It had resulted in the dispersal of its forces, and had set up tiny cells, introducing into many kolkhozes alien elements which were met with general hostility. None of this provided the training which the countryside needed. Moreover, the need of the countryside for communists clashed with the Party's general recruitment policy in the post-war period. Immediately after the victory, the Party tried to open its ranks 'to the best', and the part played by the intelligentsia in the new engagements was very important at that time. After 1949, it concentrated on broadening its social base, but it was the workers who profited from this. The recruitment of the peasants remained stagnant; this was not where a solution to the problem of training could be expected, but it could be found in the organisation of the structures of production which began in 1950 with the regrouping of the land under cultivation.

After 1945, the authorities tried constantly to increase agricultural production and to integrate the peasantry. Here they came up against the perennial problem of relations between individual ownership and collective ownership. Although in 1946, because no other way was open to it, the government adopted a tolerant attitude towards the free market, which, by the way, was provided for in the Five Year Plan 1946–50, its attitude stiffened after 1947. The State increased its compulsory levies, but it did not increase prices. The peasant responded by concentrating increasingly on his own plot and by neglecting the communal work. In 1948, the government stepped up the measures designed to restrict private production. The taxes affecting sales on the free market were increased and the duties on agricultural production were raised by a decree of 13 July 1948. In relation to the scale established in September 1929, the scale of 1948 introduced a rise of 8 to 11 per cent of the tax on an income of 3000 roubles, from 12 to 16 per cent for an income of 6000 roubles. This policy helped to lower the peasants' standard of living, but did little to increase his concern about the collective production. Furthermore, it changed the structure of individual production, the peasants turning to the most profitable crops: vegetables instead of cereals, and running down their livestock in order to avoid the taxes exacted by the State. The take-off of agricultural production was slowed down by the financial difficulties which hampered the farmer.

In 1949 the situation in the countryside was so worrying that the government considered making more radical changes.

The period between 1949 and 1951 was marked by arguments on the rural policy and the measures to be adopted. A. Andreyev was succeeded by another agricultural specialist, Nikita Khrushchev who, after the war, had concentrated on the problem of agriculture in the Ukraine where he was both President of the Council of Ministers and First Secretary of the Party. At the end of 1949, he was appointed secretary of the Moscow Obkom and secretary of the Central Committee. Between Khrushchev and Andreyev differences of opinion and certainly also personal differences developed. But the personal conflicts should not divert attention from the importance of the conflict which was to break out and which was to do with the way production was organised. On 19 February 1950 *Pravda* published an article against Andreyev's agricultural policy centred on a fundamental point: what was the best form of work in the kolkhozes? the 'link' (*Zveno*) or the 'brigade'?

The argument was important because it questioned the entire organisation which had previously existed. Since 1939, the Soviet government had always considered the 'link' as the active element in production. The 'link' was a small cell, a restricted group which in many cases was actually the family cell. This form of organisation was justified from two points of view, the technical and the human. On the technical plane, it suited a countryside which was insufficiently mechanised and very backward, and where individual initiative was still a decisive factor. From the human angle, when before the war the Soviet government attempted a reconciliation with the peasants, it was thought that this form of work in little groups should be adopted as being more to the peasant's advantage and also satisfying his individualism. Since 1939, the 'link' had been the foundation of the work in the kolkhozes and during the entire period of reconstruction the system had been maintained.

The article in *Pravda* stated that this set-up was based on an entirely false conception, that its main consequence was the strengthening of individualism, and of family and personal loyalties to the detriment of the collective conscience. For *Pravda*, 'linkage' enabled the peasant to escape from the collectivity and gave him a genuine refuge from it, and in the end the kolkhoz became merely a juxtaposition of small family groups or groups of friends which in no way formed a socialist collectivity.

Elsewhere, the article stressed another weakness in the system, connected with the shortage of rural cadres. Was it reasonable to try to integrate the small cells, whereas their regroupment into large brigades would liberate the cadres? Lastly, another argument, based on economic rationality, was added concerning the progress of mechanisation. Mechanisation formed the basis of the work of the large brigade; to pass to this stage was simply to adapt to the increase in the Soviet stock of agricultural machines.

The attack in *Pravda* revealed the central problem of the training of the countryside and led to the tightening of the authorities' control over the rural community. Andreyev, held responsible for the past mistakes, proceeded to the second plan and from 1950 work was organised on the basis of brigades instead of 'links'. The hostility of the peasants to this reform was profound and it probably explains in part why in the ensuing years the agricultural situation remained critical.

The next stage in the reorganisation of the countryside was the regrouping of the kolkhozes, which also began· in the spring of 1950 and which had its source in the same concern: to reassert political control over the countryside. The kolkhozes had not only been broken up during this period, but they had also proliferated. Those which had been set up after 1945 were often very small, giving rise to problems about equipment and cadres. Here again at the general level one could see the difficulties the government encountered within the kolkhozes, difficulties which led it to condemn the 'links' in favour of the brigades. Both in the regrouping of the workers and the regrouping of the kolkhozes, the meaning of the operation was the same, as was the technique used. On 8 March 1950, Khrushchev had published in *Pravda* a plan for regrouping the kolkhozes which was put into operation immediately and had spectacular results. In one year, the number of kolkhozes went down by half and this movement was to continue until 1954. In 1950, there were 252,000 kolkhozes in the USSR, 121,000 in 1951, 94,800 at the end of 1951 and about 83,000 in 1954. The reduction in the number of kolkhozes was accompanied by a parallel reduction in the peasant plot.

Finally, the government in the same period reduced the distributions in kind which until then had been an important part of the wages. The peasants were very attached to these distributions which, in their eyes, were more important than the monetary payments, because they were able to sell the surpluses at a high price on the free market. The government was able to justify this policy by the fact that the areas cultivated by the peasants individually in

1950 were two and a half times larger than in 1940. Nevertheless, the peasants resented the measure. The policy had a twofold effect: it solved the problem of integrating the peasants within the system but at the same time increased their resistance. The political integration had been practically solved thanks to the regrouping. In 1952 three-quarters of the kolkhozes had at last become communist organisations which this time were not simply embryonic, but genuine organisations with at least the minimum fifteen members demanded for the election of a bureau and a secretary. The progress was particularly evident in the lands acquired and collectivised after the war, where the Party made great strides. In Moldavia, the progress of the communists in the countryside was four times greater than that of the communists in the towns on the eve of the XIXth Congress (1952).

However, the numerical advance of the communists in the kolkhozes and in some outlying regions concealed a different situation: the total number of communists in the countryside remained static. This was probably due to the slowing down of recruitment which was characteristic of those years, and of the movement of the peasants towards the cities, especially of the more educated ones who were communists. Besides, the transformation of the structure of the work and the regrouping had had negative economic consequences, because of the peasant resistance. In some kolkhozes, the regrouping had meant that the livestock had been slaughtered and the Central Committee had to forbid this through a decree of 31 July 1950; the rebuilding of the national livestock suffered from this.

Lastly, the regroupments, economically sound in some cases, were in others far more difficult to justify. In the steppes, the creation of vast estates of 3,000 to 4,000 ha taking in many villages took place without hardship. In the poor areas, from Minsk to the High Volga, lands cut by marshes, forests and fallow land were regrouped together and farms of 4,000 ha often lacked any cohesion.

The measures which transformed the countryside aroused both resistance and argument, not only among the country people whose only way to oppose them was to adopt a passive attitude towards work, but above all in the ranks of the Party, notably among the non-Russian communists. In Azerbaizhan Baghirov, the First Secretary of the Party and one of Beriya's close collaborators, attacked the fusion of the kolkhozes which he thought had been done too quickly and with too little concern for the interests of production. Khrushchev, the direct or indirect target of the critics of the agrarian policy followed since 1950, wanted nevertheless to crown his work with a radical change in conditions of life in the countryside. On 4 March 1951 *Pravda* published his plan for *agrocities* which he had put forward in a speech in January. The agrocity was a large city in which the rural workers would lead an urban existence. Trucks would convey them to the fields in the morning, bringing them back in the evening. Forcibly removed from the *izba*, the framework of individual life, to collective buildings, far from their plots which would be situated on the outskirts of the cities, won over to an urban way of life by the collective conveniences and equipment, the peasants would thus lose their tenacious, deeply rooted sense of individual awareness and become workers similar to the others, integrated into the collectivity. The plan was attractive because it solved at one blow two problems: by changing the peasant consciousness it would abolish the peasant; it would abolish at the same time the difference between rural work and work in the cities, between peasant and worker and achieve the unity of the proletariat which was the mark of a socialist society. At the same time, the idea seemed unrealistic. How could the speeded-up construction of agrocities be thought of while the urban population still lacked decent housing? Here the resistance was so strong that the plan had to be abandoned. The day after the plan appeared, *Pravda* published a correction explaining that it had forgotten to point out that this was not a plan, but a debate.

Some days later, in the *Kommunist* of Erevan, Arutyunov, secretary of the Armenian organisation, violently attacked the project, followed in May by Baghirov who in *Bakinskii Rabochy* (26 May 1951) criticised once more Khrushchev's policy and ideas.

In this debate in 1951, Stalin remained silent. But the following year, he put forward his ideas in his last work *The Economic Problems of Socialism in the USSR*, published on the eve of the XIXth Congress (1952) by *Bolshevik* and *Pravda*. Although he did not commit himself on the problem of the agrocities, which he passed over in silence, Stalin seemed to agree with many of Khrushchev's proposals, especially those concerning the MTS. The peasants claimed for the kolkhozes the right to purchase the MTS and to possess their own equipment. This demand ran counter to Khrushchev's argument, which was that there should be an ever wider collectivisation. Here Stalin supported his position, stressing that if the kolkhozes were to acquire the MTS, they would become by so doing the owners of their means of production, which would be a retreat from collectivisation and from the communist future.

Furthermore Stalin emphasised his preference for State ownership, and considered that the kolkhoz was only a transitional structure which must be brought to an end together with all private production. Thus, the future which Stalin foresaw for the peasants reversed the policy followed since 1945, and seemed to indicate that the era of concessions had come to an end.

It is difficult to draw up the economic balance-sheet of the years of reconstruction in coherent terms. The fourth Plan (1946–51) was built on the reconstruction and development of solid economic bases for later advances which would be set out in the fifth Plan. In 1946, the planners paid no attention to the consumers; the only thing that counted was the economic capacity of the country. The targets were ambitious. On the basis of the index 100 in 1940 the plan aimed at attaining index 150 when it ended in industrial goods, and tried simply to catch up again with the level of 1940 in consumer goods.

It was not only a question of restoring what had previously existed; the plan foresaw the exploitation of new deposits, and the extension of the means of transport. The sums envisaged were large: 41.2 milliard roubles, that is one-quarter of the national income, were to be devoted to it. But the financial means were not sufficient, the execution of the plan presupposed a high output, a constant co-operation from the available manpower. The efforts demanded of the Soviet workers were also considerable.

In 1950, the results did not always correspond to the projects. Overall, the plan had been fulfilled, but its goals were often ill-defined, transport especially being inadequate to guarantee the provisioning of the factories and the movement of crops. The waste of goods and time was in fact very high. Above all, consumer goods remained below the forecasts.

The fifth Plan (1951–56) set even more ambitious goals than the fourth. It forecast a rise in industrial goods of 80 per cent (with high points; hydroelectricity: 170%; machines: 160%). Consumer goods, in very short supply, were to increase by 70 per cent. However, in 1952 it was clear that the targets for consumer goods and agriculture would not be reached. The entire effort once again was concentrated on heavy industry. The population found it increasingly hard to suffer from a policy which only improved its standard of living very slowly. The Soviet worker, who in 1950 shivered in his threadbare clothes, was indifferent towards the grandiose schemes for damming the Pacific in order to deflect the cold currents from the Siberian coastline. The urban population, exhausted by the years of war and the privations, found it very difficult to give what was demanded of it. The fall in the output of the industrial worker was frequently denounced by the authorities at this time; it was largely due to physical fatigue and loss of heart. The improvements after 1947 due to the government's policy were too restricted to win the loyalty of the working class.

In the countryside, the harsh attitude of the government drove the peasants back into their pre-war inertia. Seeing their material advantages diminish, threatened after 1950 by a greater integration, the peasant once again became the 'enemy' of the town, the one who was always suspected of being responsible for the poor provisioning and the high prices.

The economic recovery of the USSR, unevenly spread among the sectors, but undeniable, was thus accompanied by a double phenomenon at the base. The masses drew away from the government to which for a time during the war they had come closer; the city and the countryside were once again divided and opposed. The political consequences of this evolution were to appear all too clearly during the 1950s.

Chapter seven
THE RECONSTRUCTION OF THE IDEOLOGY

The war not only caused serious damage to the economic structures of the USSR, it also partly destroyed the ideological monolithism which had been constantly growing since the beginning of the 1930s. Among the nations and the intellectuals the liberalisation of the war period brought about a growth of centrifugal tendencies, of free movements of ideas. In both cases, the government was forced to reassert its control over the minds of the people by bringing into force successive measures which by 1951 had restored the former monolithism. Ideological unity remained after the war the fundamental postulate of the Soviet Union.

THE INTEGRATION OF THE NATIONALITIES

The war, as we have seen, encouraged the rebirth of national sentiments in spite of the pre-eminent place assigned to the Russian people in the definition of the nation and of Soviet patriotism. In fact, during the war, Stalin's attitude was contradictory, encouraging simultaneously the patriotism of the non-Russian nationalities, and the national patriotism of Russia. As the issue became clearer, he insisted more on Russia and in 1945 he definitively came down on the side of Russia. In the victory speech which he delivered on 24 May 1945, Stalin proclaimed that the Russian people had played the leading part in the war and had been the main architect of success. He then toasted not the Soviet people but the Russian people and gave three reasons for doing so: 'Russia is the leading nation of the Soviet Union.' 'Russia in the war had won the right to be recognised as the guide of the whole Union', and lastly the Russian people was marked by 'lucidity, strength of character and patience'.

This speech was of considerable importance, not only theoretically: its practical consequences were soon to appear. In theory, Stalin by this toast changed the hitherto accepted idea of the Russian people as *primus inter pares* and of the mutual enrichment of the peoples, and substituted for it the pre-revolutionary vision of the Russians as civilisers, masters and protectors of the other peoples. In so doing he rejected the theory held in Lenin's time and opted for the practice followed since 1924. The Soviet Union had always been divided in its system of relations between nations between an egalitarian theory and a far less egalitarian practice. But until 1945 no one had questioned the theoretical foundations of the nationality policy. In doing so, Stalin was true to the ideas he had defended in 1922 against Lenin in his plan for a federation, in which Russia was to be the centre of the system. Faced with Lenin's hostility he had renounced the idea; but in 1945 he inserted it into the texts.

This Stalinist trend was bound to cause a crisis among the non-Russian peoples. But even before the end of the war he started to assimilate the nations and destroy the national aspirations which were finding expression, by repression, by a historical and cultural revision, and also by the political integration of the nations. These were the essential aspects of the national policy which he carried out after the war.

Repression was the first method used to intimidate the nationalities and reduce the excessive national tendencies. Theoretically, the repression was concerned with the way collaboration could be put into practice: in practice it had a far wider significance. During the war, Stalin, referring to the problem of the attitude of the nations to the German invasion, had declared that if collaboration were to be punished, the whole Ukrainian people would have to be deported. It was clear that the government could not, without causing serious unrest, deport nearly 18 per cent of the population of the USSR; this is why they made an example of the insignificant nations. However, the system had to retain the penalties levied on whole nations and not on individuals or isolated groups. On the other hand, the system stressed the heroism of the whole Russian people. The first national group punished in this way was that of the Volga Germans. A decree of 28 August 1941, referring to the existence of this group of 'deviationists' and 'spies' announced that they were to be transferred to other regions. During the First World War, the Tsarist authorities had also envisaged the deportation of the Volga Germans and the *Large Soviet Encyclopedia* in its first edition had inveighed against those 'barbarous plans'. During the war, as the occupied territories were liberated, other peoples were affected. Tatars of the Crimea, Chechens, Ingush, Kalmuks, Karachais and Balkars were deported to Central Asia or to Siberia between October 1943 and June 1944. In all, seven nations were uprooted from their homes. This represented at least 1 million people (in 1939 there were 407,690 Chechens, 92,074 Ingush, 75,737 Karachais, 42,666 Balkars, 134,271 Kalmuks, 380,000 Volga Germans and more than 200,000 Tatars of the Crimea.

A decree published in *Izvestia* on 26 June 1946 announced simultaneously the deportation for crimes of treason of the Chechens, Ingush and Tatars and the suppression of the Checheno-Ingush Autonomous Republic as well as the transformation of the Autonomous Republic of the Crimea into the region of the Crimea. For nearly ten years the deported nations ceased to have any official existence; the *Large Soviet Encyclopedia* in the volume on the USSR in 1947 does not mention them among the national groups living in the USSR and they had no representatives in the Soviet of the nationalities after the war.

In the Baltic States, partial deportation of those who had been in opposition took place after the war; the grounds for this were not collaboration but resistance to Sovietisation and collectivisation. In order to subdue the peasants and the armed groups of partisans, the government deported in 1948–49 about 400,000 Lithuanians, 150,000 Latvians, 35,000 Estonians. In the Ukraine the armed groups of partisans had also to be subdued. In 1944, the Soviet government granted an amnesty – an exception in the history of this epoch – to 'the Ukrainian insurrectional army' but many of its members remained in groups; these were the Banderovsty (troups of Stephen Bandera); they operated within the Polish frontiers demanding the annexation by the Soviet Ukraine of the Polish territories, populated by Ukrainians, of Kholm and Khrubeshov. The 'pacification' of the Ukraine, complicated by agrarian problems, was not achieved until 1950.

Even before the 'transfers of nations or groups' had been completed, the government launched into another battle against the historical and cultural foundations of the national sentiments of the non-Russian peoples. Here again, the mould in which the Soviet nation was to be cast was no longer to be made up of the common capital of all the peoples, but was to be essentially that of the Russian people.

The rewriting of history went through a new stage in which,

once again, the past relations between the peoples of the USSR were defined. This revision was essentially a glorification of the Russian past and the problem of the ethnogenesis and origins of the Russian State were placed at the heart of the national pride. The historian Grekov stated that in the ninth century the Russia of Kiev was as developed as the Carolingian State. Pan-Slavism also gained ground within Russia. The common origin of the three Slav peoples was insisted on: Russian, Ukrainian and Byelorussian, all three having emerged from the old Russian State. The revisions of the more recent Russian past were equally sweeping and Stalin even corrected Engels, who, according to him was wrong when he stated in his article 'The foreign policy of Tsarist Russia', that 'Russia was the last bastion of European reaction'. On the contrary declared Stalin, 'the last bastion of reaction at the end of the nineteenth century was no longer Russia but the bourgeois imperialist States of Western Europe'. Congratulated by Soviet historians for this 'remarkable example of a creative Marxist approach', Stalin had brought about an important change in Soviet ideology, the affirmation of the permanent virtues of Russia. 'Socialism in one country' was justified by this insistence on the progressive character of Russian history.

The importance given at the time to historical problems is shown by the Party's weighty intervention in the writing of history. *Istoricheskii Zhurnal*, which was thought to be an unsuitable vehicle for the historical approach, was suppressed, and a new journal, *Voprosy Istorii* (Questions of history) was started; Party dignatories who were far from being historians – Zhdanov, Alexandrov, Morozov, Litvin – churned out directives, warnings and indictments. Even the most firmly established historians – Pankratova, Bakhrushin, Rubinstein – found themselves in difficulties. Two subjects were constantly insisted on in the Party directives to the historians: the influence of Kievian Russian on Western Europe must not be underestimated and the unity of the historic process of the peoples of the USSR must be stressed. The peoples of the USSR had no separate histories, but shared the same history at the heart of which lay Russia. Seen from this angle, the national past of the non-Russian peoples was decisively changed, a change which had been foreshadowed before the war but the extent of which was astonishing. 'Bourgeois nationalism' was denounced everywhere. Its manifestations could be very varied; for the Ukrainians it meant to forget their fundamental community with the Russians – ethnic, historic, economic and social community; for the non-Slav peoples it was to exalt the past before their entry into the Russian Empire as a golden age.

Two traits characterised the evolution imposed at that time upon the national histories. In the first place, the pre-revolutionary history of these peoples was presented in the light of a growing friendship between themselves and the Russian people. It was admitted that the Tsarist bureaucracy had not been a beneficial factor, but the annexations were presented more and more as progressive factors imposed by numerous external causes and almost wiped out by the *growing together of the peoples* and its ultimate result, the Revolution. The other characteristic was the clear-cut condemnation of the movements of the national resistance to Tsarism which, in the years 1945–50 affected all the national heroes, regardless of the content of their movements. Two great campaigns of revision marked that period: in Kazakhstan during the three years 1947–50, the Party tried to force the most important of the national historians Bekmankhanov to adopt its views. Defeated, Bekmankhanov preferred to stop writing rather than to go back on his work and his past.

In the Caucasus it was around the Iman Shamil that the Party tried to put its ideas into force. Here again, the Party clashed with the national historians who, behind their attachment to the national heroes, defended the past of their nation. In 1951 everywhere, the condemnations of the Party had eliminated any who were recalcitrant. The true national heroes recognised after this period were those who linked the destiny of their country to the destiny of Russia and in 1950 the secretary of the organisation of the Kabarda Party named them: Stalin, Kirov and Ordzhonikidze.

Deprived of their historical past, the non-Russian peoples were forced to identify themselves historically with the Russian people, their 'elder brother'. The conclusion of the revision of history which took place between 1946 and 1952 was presented to them by one of their own people, Baghirov, the leading communist in Azerbaizhan who wrote: 'The directing force which unites, cements and guides the peoples of our country is our elder brother, the great Russian people ... The Russian people because of its virtues, earns the confidence the respect and the love of all the other peoples.' It was also to Baghirov that we owe in 1952 an unequivocal appreciation of the meaning of the Tsarist conquest. 'Without underestimating the reactionary nature of the colonial regime of Tsarism, it must not be forgotten ... that the annexation of the peoples by Russia was for them the only way out and had a uniquely happy influence on their future destiny.' Thus, thirty-five years after the Revolution which had wanted to make a clean sweep of the past, Soviet ideology had come full circle. Tsarist colonialism seen at first as an absolute evil, then as the lesser evil, was recognised as an absolute good for the conquered peoples.

Having been deprived of their history, the non-Russian peoples were also, after 1951, dispossessed of their cultures which were denounced as elements of national differentiation, turning the peoples towards the past, and separating them from their elder brother. Here, the attack on the cultural plane was directed primarily against the Muslim peoples, who were suspected of being attached, through their common culture, to a common universe, alien to Russia, that of Islam. The Stalinist compromise on the culture of the peoples of the USSR 'proletarian in content and national in form' was, especially in the war years, understood by the non-Russians in its second form and the exaltation of the national forms was undeniable. In 1951, the Soviet government subjected the national epics of the Muslims, symbols of the national cultures, to a systematic criticism and ordered them to be forbidden. The attack began in the spring of 1951 on the Azeriah epic *Dede Korkut* which retraced the history of the Oghuzes. It was condemned for its 'clerical, pan-Turk and anti-popular' tendencies. In the summer of the same year, it was the turn of the Turkomar epic *Korkut Ata*, a local variant of *Dede Korkut*. At the beginning of 1952, the Uzbeks were ordered to ban *Alpamysh* which praised the struggle of the Turkish Kungrats against the Buddhist Kalmuks, and the Kazakhs were deprived of their whole epic cycle: *Er Sain, Shora Batyr* and *Koblandy Batyr*. Shortly afterwards, it was the turn of the Kirghiz whose epic *Manas* retraced the struggle of the Muslim nomads against the Kalmuks, at that time also called Chinese.

Although in most of the Muslim republics the national intelligentsia accepted in silence this rape of their national values, in the little Kirghiz republic this was not the case, and the resistance came even from the ranks of the Party. The argument developed with exceptional violence for months on end in the official bodies of the central committee of the Kirghiz Party, the Russian publication *Sovetskaya Kirghiziya* accusing the natives of bourgeois nationalism, the publication in the Kirghiz language *Kyzyk Kyrghizistan* (Red Kirghiz) replying that the Russians crushed the cultural values of the other peoples in order to impose their own. For months, it was necessary to mobilise most of the scientific and political, local but also central, authorities, to move the cadres in order to break the resistance of the intellectuals. But the crisis had shown the profound effects of the concessions made during the war. Under pressure, the intellectuals generally submitted in silence, refusing to recant, and during the struggle the grievances which they expressed were a sign of the renaissance of the national values.

The bringing to heel, in these conditions, had to be political also within the local parties. Although in most of the republics, some movement in the cadres marked the years of reconstruction, one of them – Georgia – underwent an exceptional political crisis. On 1 April 1952, the Plenum of the Central Committee of the Party, convoked in the presence of Beriya who had come especially from Moscow, sacked the entire leadership which had held office since 1938. The Minister of the Interior Rapava suffered the same fate. The new leadership headed by A. Mgueladze who undertook between April 1952 and March 1953 a very wide purge which affected not only the Party (427 secretaries from Gorkom and Raikom) but also to a lesser extent the State apparatus. The Georgian press remained practically silent about the reasons for the purges, referring incidentally to the lapses in discipline and to the 'bourgeois nationalism' which had been discovered everywhere. At the XVth Congress of the Georgian Communist Party, the new First Secretary denounced the mistakes of his predecessor, Charkviyany, and his speech was an astonishing mixture of submission to the general directives of the Communist Party of the USSR in the 1950s, 'struggle against the national tendencies' and the defence of the Georgian interests against the national minorities of Georgia–Adjars, Ossetes, Abkhazes – over whom the new leadership tried to re-establish Georgian authority. The 1952 crisis in Georgia revealed, like the Kirghiz resistance around the *Manas*, both the Stalinist determination to crush the national sentiments of the non-Russians and the strength of the national feelings. But the national policy of the Soviet government was not limited to these negative aspects. The USSR had learnt the lesson of the war and knew that nations could be destroyed by force. The constitutional texts and the Party attempted, together with the repressive measures, to integrate the nations by increasing their participation through the cadres, some of which were Russified, and some which were not.

The Soviet constitution was changed in 1946 in order to justify the entry of the Ukraine and of Byelorussia into the United Nations, together with the USSR. These modifications meant for the federated republics the right to have diplomatic representatives and national armed forces. These were theoretical rights, no doubt, but they strengthened the national convictions, the feeling of difference, rather than of units. Above all, the Party tried to diminish the excessive degree of Russification which had characterised it at the beginning of the war and to introduce more non-Russians into the local organisations. Of course, the Party hoped that the national communists would be the instruments of a growing 'internationalism' which in the post-war years was taken to mean Russification. Whatever were the intentions behind the changes, Table 3 shows that at the time of the XIXth Congress the position of the national groups in the Party had improved for the nations of Central Asia, Georgia and Armenia, but had remained unfavourable for the Ukraine, Byelorussia and the Baltic States. The following years saw a slight progress of nearly all the national groups, but in 1952 the local situations were still very marked by the legacy of the war. The small share in the Party played by the Ukrainians can be explained by the situation which had existed in 1940, the enlargement

Table 3.   Position of the national groups in the Party in 1952

| | Percentage of the population of the USSR | | Percentage of delegates to the Congress | |
Republics	1939	1959	XVIII (1939)	XIX (1952)
RSFSR	63.9	55.9	65.8	65
Ukraine	18.2	20	18	12.8
Byelorussia	3.3	4.1	2.9	2.2
Georgia	2.1	1.9	2.5	2.7
Armenia	0.8	0.9	1	1.1
Uzbekistan	3.7	3.9	1.5	2.1
Turkmenia	0.7	0.7	0.4	0.7
Kirghiz	0.9	1	0.3	0.7
Tadzhikistan	0.9	0.9	0.3	0.5
Kazakhstan	3.6	4.5	2.5	3.5
Estonia	–	0.6	–	0.5
Latvia	–	1	–	0.8
Lithuania	–	1.3	–	0.6
Moldavia	–	1.4	–	0.3
Karelia	–	–	–	–

also of the Ukrainian republic, which had been increased by a non-communist population. This was also the case in the Baltic States.

Although after the war a slight progress in the non-Russian participation in government was registered, a careful examination of the fundamental posts in which real power was concentrated – government and secretariat of the cadres at Party level, reveals that the situation existing before the war had not changed. Everywhere, the key posts were held by Russians and the national cadres were restricted to positions of local concern with limited responsibilities. In spite of the appeals for the 'indigenisation of power' the national policy seemed, in the end, to favour the Stalinist conception of supervision by the elder brother. On the plane of relations between the nations, the policy carried out after 1945 undoubtedly re-established the former order, put an end to all the concessions made during the war and stamped on the evolution of the USSR a more Russian colour than before 1940. In 1952, the nations seemed to accept this unifying vision and the calling to order had apparently come to an end.

THE ZHDANOVSHCHINA

The liberal tendencies had not only marked the evolution of the nations; they were also characteristic of the entire intellectual life of the USSR. Immediately after the war, the Soviet intelligentsia tried to shake off the shackles which had constricted it for so long. There were various signs of this new intellectual independence after the war. The publication of the *Recollections* of Serge Alliluyev, the old Bolshevik worker who had hidden Stalin many times in his home before the Revolution and then became his father-in-law was typical in this respect. The picture presented by Alleluyev of the pre-revolutionary years relegated Stalin to the background, and differed from the official history presented by the Party in 1938, which has since become the supreme reference. The same was true of the youthful recollections of Stalin's former sister-in-law, published at that time in *Pravda* in which Stalin was treated very cavalierly, very different from the legend enshrined by the Party.

But after the summer of 1946 the situation changed suddenly; for the intelligentsia the time for being brought to heel had come. The reasons behind the hardening of the government's attitude were obvious. In attacking piece by piece the Stalinist legend or the history of the Party, the intelligentsia was moving little by little towards the destruction of the body of established beliefs. How could a line of demarcation be drawn between questions which were harmless and those which were not? How could the criticism be stemmed when it attacked fundamental problems? Could not the revision of Stalin's role in the Revolution lead to a reflection on his role after 1924? To a revision of the whole history of the Party? Stalin saw clearly that truth is not made up of isolated pieces but is a whole; the revision of one aspect leads inevitably to the whole being challenged.

The refusal to allow criticism and research to expand, was coupled with the government's concern about the attraction the West had for the USSR. This attraction was incompatible with the glorification of Russia in which the government was engaged at the time. Although the reasons for the disciplining of the intellectuals are clear, the reason for the moment chosen to unleash the attack against them is less so. One suggestion can however be put forward: the economic crisis of the summer of 1946, due to the drought, compelled the government to reduce the intelligentsia to silence.

Faced with the famine, the criticism of the intellectuals was inevitably directed towards the whole economic policy and could have led to a comparison between the advantages of planning and those of Western methods. In the USSR in 1946 the intellectuals were no longer disposed to believe blindly that a modern nation could be reduced to starvation because of a drought. Thus the economic crisis seems to have precipitated the attack against the intelligentsia which was to be three-pronged: an attack against formalism and cosmopolitanism, intervention in the universities, and economic debate.

The first sign of the change was the announcement of the creation on 1 August 1946 of a new journal *Partiinaya Zhizn* (the life of the Party) which according to its first editorial was to have the task of calling on the Party to be vigilant about the general trend towards ideological weakness and depoliticisation. The journal was to focus on the intellectual, scientific and artistic life of the USSR. It appeared thus, from the beginning, as an organ of control, in a country which was trying to forget controls. In the following days, several decrees completed the devices by which intellectual life could be supervised. On 2 August 1946 a decree reformed the Party education system in order to fit it for the fight against 'foreign influences and the new ideas which were weakening the communist spirit'; it implicitly foreshadowed that the declining intelligentsia was being relieved. Some days later, the Party passed to the attack against the most damaging tendencies, formalism or again aestheticism, and demanded that everywhere the *Party spirit* (*Partiinost*), which alone conformed to Marxist-Leninist ideology should take their place. The idea of ideological unity, of the ideological content of intellectual life was thus reintroduced. In the USSR, nothing could take place outside Marxism-Leninism as defined by the Party. Two great literary publications in Leningrad, *Zveda* (The Star) and *Leningrad* were denounced on 14 August for having given prominence to ideologies alien to the Party and although the former was able to survive for a time with a new editorial board, the second was suppressed.

Behind the general reasons, individuals were increasingly censored. The poetess Akhmatova and the satirical novelist Zoshchenko were excluded from the Union of Writers. Writers, painters, musicians, film producers were denounced one after another. The novelist Platonov was accused of having libelled the Soviet railways in the *Ivanov Family*. Zhdanov, who was promoted to the rank of cultural leader because, wrote Stalin's daughter Svetlana Alliluyeva, 'my father had the impression, goodness knows from where, that Zhdanov was competent in cultural matters,' one day in the summer of 1946 summoned Prokofiev and Shostakovich to give them at the piano a 'lesson in communist music'. His dumbfounded listeners wondered if they were dealing with a madman. But Zhdanov was to have the dubious honour of bequeathing his name to this period of intellectual dictatorship which continued long after his death in 1948. On 10 February 1948 a decree on 'the decadent tendencies' of Soviet music condemned simultaneously the Georgian opera of Muradeli *The Great Friendship* and the 'anti-popular' formalism of Prokofiev, Shostakovich, Khaychaturian and other musicians. The anti-formalist campaign was pursued until the beginning of 1949 with a very radical result: nearly all the well-known intellectuals and artists were denounced, purged or forced to restrict their activities. Most of them took refuge in silence and the intellectual and artistic production of the USSR, at least in official circles, fell to a very low level.

From 1949 onwards, the denunciation of formalist tendencies was eclipsed by the discovery of a new deviation, cosmopolitanism. The anti-cosmopolitan campaign began with a refusal to allow the 'servile cult of the West' to take root in the USSR and turned into a disguised but nevertheless real anti-Semitism. The first anti-cosmopolitan move was the banning of marriages between Russians and foreigners. The USSR retreated into itself and withdrew behind its frontiers. At the end of 1948, the denunciation of cosmopolitanism took a precise turn. The Jewish Anti-Fascist Committee created during the war was banned; nearly all its members were arrested; its president, the actor Mikhaels was killed in suspicious circumstances. Svetlana Alliluyeva stated that he was killed with the consent of her father. The newspapers were full of revelations about the 'real' name of various cosmopolitans. The campaign was so violent that in February 1949 Stalin ordered the newspaper to put a stop to the 'denunciation of literary pseudonyms which had an accent of anti-Semitism'. But 'rootless anti-cosmopolitanism' remained under fire by the Party.

This first campaign against the intelligentsia was completely dominated by the desire to drag it away from the attraction of the world outside Russia and to root it in the genius of Russia, which was the leader not only of the peoples of the USSR, but also of Eastern Europe. The intelligentsia's desire to open itself to the outside world compromised the balance which Stalin was trying to reach. By the end of his campaign he had succeeded in forcing all those who after the war expressed intellectual curiosity to remain silent, although he had possibly not been able to force them to accept his ideas. This rejection of the world outside was confirmed at the popular level by the signs which the government erected at the entry to the villages on the return of the prisoners of war. Proudly, the signs proclaimed that the talk of those who had been out of the USSR should not be believed because the Soviet situation was far better than the situation in the West. If they dared to say otherwise, they were lying and were traitors to their country. The Soviet government had to defend itself against the reminiscences of the demobilised soldiers, as well as against the imagination of its intellectuals. Both bore the stamp of a truth which the Party denied.

Simultaneously with this campaign against the writers and the artists, the government began to attack the university teachers who, like the whole Soviet population, tended to think for themselves and not according to the official directives. The Party laid down the law in all fields – philosophy, linguistics, biology, mathematics – and indiscriminately condemned wave mechanics, cybernetics, psychoanalysis and many other sciences which it labelled as bourgeois. Those who defended them were dismissed, deprived of their livelihood and sometimes deported. Among the disputes, two are particularly significant, that of the linguists and that of the biologists. In the first case, Stalin intervened personally in 1950, writing an article in *Pravda* directed against the theories of Marr, although he had been dead for fifteen years. The need for this intervention, which did not attract much attention, was not clear; but by eliminating from the universities the intellectuals called 'Marrists', it made room for a new generation over whom the Party had more control. Thus, as Leonard Schapiro has noted, the purges were destined to deprive the university teachers of their feeling of security and of independence. At all levels, Soviet society was brought to a state of insecurity which inhibited the slightest criticism.

The debate over biology had a quite different significance. It was linked to the economic difficulties of the USSR. Here again, the Party took a vigorous stand against Mendel's biological theories and upheld the botanist Lysenko, whose voluntarist theories, which he was to extend to biology of which he was completely ignorant, enabled the government to state that Soviet agriculture was on the verge of a spectacular development. The only serious consequence of this unequal debate was the dismissal, sometimes the liquidation, of eminent scholars like the biologist Vavilov and the lagging behind in science of the USSR because all research based on the condemned theories was halted. Although this time-lag was later overcome, this was because many scholars continued their work clandestinely, while publicly toeing the Party line. In this chapter mention must also be made of the disciplining of the historians which was carried out on the theme of international relations. Everywhere purges which were for the most part administrative, recast scientific thinking and scholarship.

The economic debate which also began in 1946 played a special part in the control which was then being re-established over the intelligentsia. Here, it was not a question of dismissing men and teams nor of defending imaginative theories, but more profoundly of justifying by means of the economic theories the political options of the Soviet Union.

This debate exploded on the occasion of the publication of the book which the Hungarian-born economist Varga had written on economic and thus political problems of the capitalist world after the Second World War: *The Economic Transformations of Capitalism at the End of the Second World War*. This book appeared in 1946 and from the first the Party had genuine reservations about it; it led a great debate in 1947 which was published in the journal *World Economics and World Policy*. The upshot was the closing down of the Institute of World Politics and Economics directed by Varga. Starting with the analysis of the capitalist world as it appeared after the war, Varga rejected the theory of the imminent crisis of capitalism which for decades had been the foundation of the whole internal and external policy of the USSR. On the contrary, he discerned profound changes in capitalism, which had shown itself to be far more stable than had been believed and gave signs of an ability to change which communist theory could not disregard. For Varga, the post-war capitalist State was capable of promoting decisive structural changes, without unleashing a profound political crisis. Moreover, Varga considered that the trends in capitalist society, allied to a new world equilibrium, no longer rendered conflict between the two systems inevitable, so that the interests of the Soviet Union lay in co-operation with a changed Western world, which was evolving unobtrusively and not in the pursuit of a hostility which weakened the USSR itself.

Varga's theses had considerable effects upon the internal and external policy of the USSR. In foreign policy, it meant that the offensive attitude of the years 1945–47 (marked by the advance of socialism in Eastern Europe and the revolutionary approaches in the north of Iran) was abandoned, and later the renunciation of the 'cold war' which had begun in 1947 and had been marked by the extension of the notion of the besieged fortress to the whole European socialist camp, and the beginning of a policy of co-operation. This was precisely what the Soviet government had eschewed in 1947; it was against such a revision that it had forced its allies in Eastern Europe to reject the Marshall Plan and to align themselves with the Soviet political system; it was against this, that the Soviets had recreated, in the Cominform, an instrument which unified the parties in the name of future revolutionary tasks.

On the internal plane, Varga's ideas led to a no less radical break with the Stalinist way of thinking. The authoritarian political system as well as the economic system were founded on the need to fight against the capitalist encirclement. To admit that this encirclement no longer existed was to admit the possibility of internal change, the rejection of authority, the promoting of the private interests of the citizens and not the defence of the collectivity. In 1947, the debate between the economists took place with a firm determination to minimise its political implications. The adversaries argued over three questions: the nature of the State at the time of the development of monopolistic capitalism; the capacities of this State to act, in certain conditions, and notably in the case of a war economy, against the interests of certain monopolies; the definition of the decisive force in a war economy, the State or the monopolies. In this way the central problem was avoided.

In spite of this, Varga was forced to capitulate, but without endangering his personal position. But his capitulation was qualified, and sixteen years later in his last book, *Essay on the Problems of the Capitalist Political Economy*, he returned to the question in order to emphasise that the question for him was not so much of recognising his faults as of preventing the 'capitalist press' from using his positions to present them as fundamentally revisionist and Western.

The political, and no longer technical, response to Varga's book came from Stalin in 1952 in his last work *The Economic Problems of Socialism in the USSR*, in which, while coming down against Varga, he allowed some openings towards the future to be glimpsed. Stalin maintained once again the thesis of the imminent crisis of capitalism; at the same time, his book showed that he himself was no longer unaware of the changes which had taken place in the world. He expressed this reappraisal in his considerations on the future of international relations in which he suggested, like Varga, that the inter-system conflicts were perhaps not the most important nor the most dangerous, and that it was within capitalism that the dangers of conflict were greatest. However, Stalin refused to link his view of the international evolution and the internal evolution of the USSR. This was first of all, because he did not start, like Varga, from the stabilisation of capitalism but from the traditional idea that capitalism was declining and that its very weakness made it dangerous and aggressive. This is why he maintained his vision of an unchanged internal policy. The two fundamental elements of Soviet policy remained for him the continued uninterrupted development of heavy industry and the acceleration of the process of transformation of rural society to reach as quickly as possible the suppression of every form of private ownership. Although, like Varga, he accepted the idea that the USSR could possibly be given a respite internationally, unlike Varga he concluded from this that this respite should be used to advance with giant strides along the road to communism. Before the publication of the book Stalin had, in April, May and September 1952 argued in *Pravda* with the economists Sanina and Venger on this theme, pointing out that the tendency to oppose the search for an economic rationality made possible by the transformation of the international conditions to the elaborate centralism of the Soviet State, was a mistaken way of looking at the problem. The rationality of the socialist State was for Stalin in 1952, as in 1936, the development of an ever more centralised State.

The reaffirmation of the necessity for the State was one of Stalin's constant concerns, and it is in this context that his intervention in the debate on linguistics must be seen. In his *Marxism and the Problems of Linguistics* Stalin analysed, by using the problem of language, the problem of the real relations of the superstructure and the infrastructure in Soviet society, which led him to reaffirm the permanence of the State. On 2 August, he returned to the problem in a 'Letter to Comrade Kholopov' which was published in *Pravda*.

In this letter he stated, this time trenchantly, that Engels' thesis expressed in the *Anti-Dühring* on the withering away of the State after the Revolution was false, 'given the capitalist encirclement while the socialist revolution has only triumphed in one country and capitalism dominates all the others. The country of the victorious Revolution must not weaken but most consolidate as much as possible its State, the organs of its State education and its army if that country is not to be crushed by the capitalist encirclement. . . .'

These articles and the book by Stalin are a sign of how the debate was persisting, although it had been officially closed by the condemnation of Varga's ideas; it was a sign also of the economists' reluctance to accept Stalin's conclusions. But in the last resort, the ideological choice of the Party, represented by Stalin, took no account of these reservations and the work of Stalin showed that the Soviet ideology had not changed fundamentally, that argument was once again impossible and that the intelligentsia must acquiesce. After the years of semi-liberty, vigilance once again became the central theme of Soviet policy, and this vigilance made it impossible for the intellectuals to publish any personal theories which infringed the Party line. Here again monolithism was gradually re-established.

THE REINTEGRATION OF THE ARMY INTO THE NATION

During the period of ideological reconstruction, the government was faced with a problem which was a potential rather than an actual danger: this was to prevent the army, surrounded with the aura of the prestige of victory, from becoming an autonomous social body which could channel the desire for change or at least could act as a counterweight to the government, and at times could be an arbiter in the debates. In 1945 the government's disquiet about the army was based on two facts. First of all, since the end of the war, the army had enjoyed considerable prestige and its leaders, especially Zhukov, who had not only defended Moscow but had taken Berlin, were immensely popular with the army and in the country – as popular as the political leaders, even Stalin. Also, the situation in 1945 was strangely similar to that of 1825. For years the USSR, like the old Empire, had lived withdrawn within itself, without contacts with the outside world, and its citizens could only imagine this world in terms of their own experience. For the first time since 1917 the masses had come into contact with the world outside, with its ideas and with its social situation. In 1815, the Imperial army had discovered revolutionary ideas and become aware of the autocracy and the backwardness of Russia, and this had given birth in France to the Decembrist movement of 1825. Haunted by the lessons of history, the leaders of the USSR probably dreaded this precedent, foresaw that the army could become a centre for the fermentation of ideas which might endanger the whole system.

Here the real nature of the Stalinist fears must be made clear. It was not the idea of a military *putsch* that was to be feared, since Russian history had no tradition of military government and the army had never played an important political role. In contrast, although it had never seized power, it had on various occasions, in 1825 and 1917, helped to change the political system and contributed towards the change. But whether as an instrument of change or as an autonomous force, manipulating the change to its own advantage, an army which had thrown off the ideological shackles of the system posed an equally grave threat. The army was brought to heel quickly enough to avoid a crisis.

Three methods were used: the integration of the army into the Party, the retirement of the military leaders and the depersonalisation of the history of the war. The integration of the army into the Party was characteristic of the years 1943–45. On the eve of the invasion in 1940, the Party had numbered 3,399,975 members (1,982,743 full members and 1,417,232 candidate members). In 1945 after the victory it had 5,760,369 (3,965,530 full members and 1,794,839 candidate members). This growth in numbers was all the more considerable in that a great many communists had disappeared during the war. It is thought that a maximum of 2 million of the pre-war Party members survived, which meant that the Party had recruited nearly 4 million new members in that period. Among the new recruits, the most important group was that of the armed forces enrolled in the Party while they were serving; this recruitment is estimated at around 2½ million, which was nearly 40 per cent of the post-war Party membership. This massive recruitment had been decided upon in August 1941 by the Central Committee which sent out a directive recommending that 'all those who had distinguished themselves on the battlefield should be encouraged to join the Party'. It was only in October 1944, with the approach of victory, that the Central Committee called upon the military organisations to be more careful about the recruitment and to take into account the political qualities of the candidates.

This new trend was not restricted to the army, but also affected the civilian recruits who were the subject of similar directives. The Party began to be uneasy about the influx of communists who had undergone no preliminary education, who were badly informed about Marxism and whose personal virtues had often been taken for granted. This concern was all the more justified in that out of the 2 million communists from the pre-war Party (barely one-third of the whole Party), half had been hastily recruited after the purges, which augured ill for the cohesion and ideological capacities of the new Party. The influence of the military elements in the Party had two contradictory results: the first was that it enabled the Party to integrate the army effectively and the second that the army could influence trends within the Party. This second consequence made it essential in 1945 to integrate the army ideologically.

The elimination from the public eye of the military leaders was also a consequence of this concern. Zhukov or any of the other well-known leaders could in fact, in 1945, rely on the support of 40 per cent of the Party members. This inevitably led to their elimination, which began in 1945, when they were forced out of public life, not by censuring them, which would not have been tolerated, but by giving them commands either in positions or in the provinces which removed them from politics. Every military ceremony, every anniversary showed how their position had diminished. Zhukov, the most popular of all, no longer appeared in public in his own country after 1946. On 4 May 1948 *Pravda*, commemorating the battle for Berlin, no longer mentioned his name; the credit for having drawn up the plan of battle was given to Stalin.

The depersonalisation of the history of the war accompanied the elimination of the military leaders. The leadership of the Party and the anonymous soldier took first place in the written history. The determination to restore the prestige of the Party, which had tended to be neglected during the war, was evident during these years. The statement that the Party had always been in control of events, that it had never, even in 1941, been taken by surprise, was contradicted by a remark which Stalin let slip in a moment of sincerity at the end of the war: 'Poor Soviet people, you could well ask by whom you were governed in 1941. You had every cause to be indignant, because we simply did not know what to do.'

This substitution of the whole Party for the military high command was the version of history which Stalin's successors also retained. *The History of the Communist Party*, published by the Party at the end of 1969, firmly holds to this version. Destalinising in all respects, when it comes to the war history of the war, it remains faithful to the Stalinist explanation. The chapter on the Second World War contains very few of the names of the military leaders: Zhukov's name only appears twice and is omitted from the battle for Berlin; when it is mentioned, it is always among the other names and in alphabetical order. In contrast, the heroes from the ranks, such as the popular soldier Martynov, are praised to the skies. The soldier took precedence over his leaders and above the soldier was the Party.

This historical conception created in 1947–48, coinciding with the disappearance of the military leaders, undoubtedly deprived the army of an opportunity to play its own hand. Now a body without a head, demoralised by the suspicion with which anyone who had travelled outside the frontiers of the USSR was regarded, the army acquiesced without protest in this state of affairs and very soon ceased to be a conscious social body. It did of course reappear after Stalin's death and acted as an arbiter in the conflicts between his successors. But its intervention at that time was invoked by those who feared the power of the police, which was controlled by Beriya. They asked for the help of the army which, after its intervention, returned to its position on the sidelines. The potentialities of 1945 were well and truly destroyed in the months which followed.

In the space of a few years, 1945–50, the Soviet government had gradually succeeded in reasserting its control over all the social forces to which the war had given a measure of autonomy. However, in this period the lessons of the war were not forgotten. Stalin and his colleagues remembered that the excessive repression had led to a quasi-collapse of the entire system in 1941 and that the recovery was only achieved at the cost of many concessions. This is why for several years, while still pursuing a policy of reasserting control and of reintegration, the government tried not to rush those whom they were disciplining, and to give the restored system a reassuring aspect. Although purges did take place, although arrests led to men being put to death, the terror of the pre-war years was a thing of the past and there were many who, disavowed by the Party for their ideas or their activities, Varga and other scholars, for instance, remained at liberty.

As regards the peasants and the nationalities, the two great masses which were so difficult to assimilate, the years 1945–50 were a curious mixture of readjustments and tolerance. In spite of the contradictions, centrifugal currents continued to appear, showing that Soviet society found it difficult to accept the return to monolithism and hoped that lasting changes had been won. After 1951, on the contrary, the Soviet system seemed to enter a new phase, the first signs of which were already visible – Stalinism was to follow on the heels of the reimposition of discipline. Return to a degree of terror, return to the pre-war ideological themes on a basis of personal conflict? The reasons for this slide are open to question. Within the country, in spite of immense difficulties, the economy was slowly recovering and the USSR was repairing its ruins. Abroad, the Cold War went on, without any visible modifications. There are two explanations possible, neither of which excludes the other. At home, the exhaustion of the people following the post-war hopes and strengthened by material difficulties, led the government towards the easy solution of growing control and repression. It preferred the more reassuring docility and passivity of the people to reflection and initiatives. For the masses, the post-war years, above all at the end of Stalinism, were years of silence. Everyone concentrated on living his own life as best he could and lost interest in a political system which the war had left unchanged; but the outward loyalty to the system and its values was remarkable; mental dissimulation was the rule. The terror was still restricted; it affected the summit and some pre-determined groups. In this sense, there was not a complete return to the years 1935–38; simply the system of dominating the masses by an apparatus which spoke in their name perpetuated itself.

Another explanation for the reinforcement of the repression by the authorities was the extension of the Cold War in 1950 by the Korean War. Military expenditure increased very rapidly. Stalin spoke against the growth of consumption and in favour of a more rapid integration and transformation of the peasantry. Whatever may have been responsible for the origins of the Korean war it was clear that the first post-war conflict was once again going to justify the obsessional complex of the USSR upon which its internal authoritarian system was built. Once again, the citadel of socialism was threatened and in the minds of the citizens the fear of war, understandable if one remembers the losses and the sufferings which they had endured, at once revived. The people's fears therefore encouraged acceptance of an authority which became more and more oppressive.

Chapter eight

STALINISM COMPLETED

Political life in the USSR in the years immediately after the war was not only marked by the efforts of the authorities to regain control of society, but by the evolution of power which arose out of this central concern and marked it profoundly. Several elements emerge at this level: the transformation of Stalin's personal power, the political conflicts at the top, the evolution of the internal policy of the USSR after 1950 and the perspectives suggested by this evolution.

EVOLUTION OF STALIN'S GOVERNMENT

In the years 1945–52, Stalin's character and power were more ambiguous than ever. The man who had collapsed in 1941 had been replaced by the victorious leader and by the head of the national State as well. In this period his legend pervaded the whole of Soviet life; he was seldom seen, he wrote nothing, but his legend grew inordinately. In all the socialist countries the cities were dominated by huge statues of Stalin; at the summit of the Elbruz (5,642 m) his statue bore the inscription: 'On the highest peak in Europe we have erected the statue of the greatest man of all time.' The Stalinist cult reached its apogee on his seventieth birthday in 1949. The presents sent from all over the world were put on display in museums and *Pravda* published endless lists of them, together with messages of congratulations. All virtues, all knowledge were attributed to him and he laid down the law about everything. His glory knew no bounds.

Amid this concert of praise and protestations of loyalty never had he been so alone. The accounts of those who knew him, of his daughter, of Khrushchev, of those who came near him more incidentally – like Djilas – all agree: irascible, devoured by suspicion, anxious, the ageing Stalin withdrew into himself, fled from others, fled from himself in the houses which he untiringly had built and in which he never occupied more than one modestly furnished room. Those he could bear to be with were his drinking cronies, the old members of the Party leadership whom he compelled during interminable nights to drink until they were completely exhausted. Stalin did not speak then except to mock or to threaten and his contemporaries were aware that these grotesque and terrifying evenings often ended in disgrace. Mad and bloodthirsty, said some historians, maintaining his power through the last purges. But this is not true. Stalin avoided the great purges until 1953 and the logic and coherence of his authority can be clearly discerned.

What he was trying to consolidate was the new system, *Stalinism*, his creation by which he replaced the system inherited from Lenin. Since the war, the Soviet State had genuinely been a new State in which the Bolshevik inheritance was constantly diminishing and, after 1946, Stalin set about erasing all visible traces of it. On 9 February 1946, he attacked the Leninist conception of the Party in a speech in which he declared that the only difference between Party members and those outside the Party was that the former were members of the Party and the others were not. It was only a formal distinction. Having thus destroyed the source of the Party's strength, he restored the titles which Lenin had wanted to abolish forever, because to him they represented the traditional State. On 15 March 1946 the Council of People's Commissars became the Council of Ministers; on 25 February 1947 the Red Army of the Workers and Peasants assumed the title of Armed Forces of the USSR. Finally, in 1952 at the XIXth Congress, the Party suppressed from its name the word Bolshevik and became the Communist Party, without any other adjective. Everywhere the past gave way to the post-war reality. The USSR was a Great Power, very similar to all the others.

The methods of political administration also underwent the same change. Stalin tried to concentrate power in the organs which he had created, bypassing the control of the ruling bodies elected in 1939. His personal Secretariat directed by Poskrebyshev from 1945 onwards was really the decision-making centre and through its powers dominated the Secretariat of the Central Committee. Stalin also created great confusion among the leading organs of the Party which affected all the work by increasing the number of posts and thus by overlapping responsibilities. In this way he created within the Politburo numerous commissions – of the Five, of the Six, of the Seven, etc. with vague and generally overlapping competences, and confused the situation still further by introducing external elements into these specialised commissions. This was the case with the Commission on Foreign Affairs (Commission of Six), to which he also entrusted problems of internal policy and into which he introduced Voznesensky, President of the Gosplan, transforming the Commission of Six at a stroke into the Commission of Seven. He scarcely ever called a meeting of the Central Committee and very rarely convoked the Politburo, which was never brought up to strength because Stalin, haunted by the fear of spies, systematically excluded some members, accusing them of being in the service of a foreign Power. This was the case with Voroshilov whom he barred from the Politburo under the pretext that he was a member of the British Intelligence Service!

As for the sessions themselves, whether within the Party or within the State organs, when they did take place they were tragically grotesque. Problems were not dealt with punctually; Stalin made all the decisions without even consulting his colleagues and forced them to agree. Khrushchev has left striking accounts of such meetings. To have the Five-Year Plan adopted, Stalin picked up the file and declared: 'Here is the Plan. Who is against?' Before his colleagues even had the time to reply, he added: 'No one. It is adopted.'

He extended the unbridled authority which he showed towards the political cadres of his country to the political leaders of the popular democracies. After for a while supporting coalition governments as agreed by the war-time international agreements, he set up political regimes everywhere modelled on that of the USSR, the outcome of 'revolutions from above', assisted by the Red Army and the political authority of the Kremlin. From all these regimes, he expected, as from his colleagues in the Politburo, a blind obedience and frequently gave vent to contemptuous statements: 'I have only to lift my little finger and Tito will collapse.'

The Cominform, which was created in 1947 in order to guarantee ideological coherence, reflected his new methods. It was not an international body, but an external agency of Soviet will. Everywhere, at home and abroad, contempt was perhaps the essential characteristic of Stalin's relations with communists. He was immured within his superiority by the certainty that he had the whip hand over the system and over the men who were part of it. And yet social tensions increasingly developed in the USSR and in the States formed in its image. The constant strengthening of Stalinist power, its absurd forms, told the story of these tensions and were probably a deliberate attempt to hide them

President Truman said of Stalin in 1948: 'Joe is a prisoner of the Politburo', and was inclined to hold his entourage responsible for the system. This idea, nourished by the echoes of internal quarrels, brings to the fore the problem of where power in the USSR really lay after the war. Was Stalin really the autocrat described by Khrushchev or Djilas? Or the leader of a minority within the Politburo, surrounded by groups which were forming in the light of the approaching succession? Political conflicts did in fact reappear in the post-war years and this was made easier both by Stalin's increasing years and the internal problems of the Party.

The fact that Stalin had aged was incontrovertible and confirmed by many eye-witnesses. The serene face which age had calmed masked a stranger. Was Stalin slipping into senility or was he pretending to be senile? He spoke constantly of age and fatigue and of retirement or even of the approach of death. He involved his colleagues in the idea that shortly his successor would have to be chosen. His daughter's evidence on this period is contradictory. In *Twenty Letters to a Friend*, she came down on the side of age and illness and reduced Stalin to the state of a plaything in the hands of a ferocious Beriya, who bore the real responsibility for the last turn towards terror. In her second book, *Only One Year*, she corrected this impression and laid greater stress on the appearance of old hatreds. The Stalin described in this book played with others like a cat with a mouse. In spite of the insistence on the senile pleasures of eating and drinking, Djilas and especially Khrushchev seemed in the last resort to retain the impression that Stalin was deliberately, and true to his old tactics, setting his eventual successor at loggerheads with each other.

The appearance of different tendencies among the Party cadres are more easily understood, if one reflects on its situation. For the first time since 1938 the internal evolution of the Party enabled men who saw themselves as candidates for the succession to seek the support of distinct groups within the organisation which outwardly remained as monolithic as ever. The Party which emerged from the war in 1945 bore the scars of sixteen years of successive crises. The last crisis, that of the war itself, differed from those which had gone before, because although, from 1929 to 1940, the Party had evolved in isolation from Soviet society, the crisis of the war had been shared by the whole of the USSR and to some extent had drawn the Party closer to the rest of the country.

The transformation in the Party was caused, above all, by its growing membership which had more than doubled. In a short space of time it had assimilated a large number of newcomers. But, and this was characteristic of the post-war Party, these new members were extraordinarily heterogeneous. They were composed of several distinct groups. First of all, there was the small group from the pre-war Party (about 1 million of them) who had continued to exercise their normal functions in the unoccupied territory. A second group, also from the pre-war Party, and composed of roughly 500,000 individuals, created, unlike the first, problems for the Party leadership. These were the soldiers evacuated from the occupied territory and the battle zones who, after the war, had to be reintegrated into the homeland, where new organisations had been set up. In fact, as the German troops retreated, the Party recreated organisations in the liberated territories in order to reassert its control over the population more quickly. These organisations functioned with members hastily recruited on the spot and numbered about 1 million, of whom half were women. Until 1945 these members played an important part and the sharing of positions and responsibilities with the former regional cadres did not take place without clashes.

Besides, the Party had never before given responsibilities to women. After the war, women who had held important positions were reluctant to retire. Similar tensions appeared in another group, numerically smaller but with a strong popular basis; this was the group of the clandestine leaders in the occupied territories, about 50,000 people. At the moment of the collapse, when all the apparatuses disintegrated, and the contacts with the centre were cut, a small group of men tried to keep the Party alive, to take over in its name the burgeoning resistance. Closely linked to the population, whose sacrifices they shared, almost independent of the central authorities, in 1945 they had to learn discipline once again and to accept the authority of leaders whose political lack of success in 1941 had been clear to them. Lastly, the armed forces must be added to all these groups.

How could such disparate elements be integrated? How could the differences between them be smoothed out in order to weld them into one single community? How could they all be made to accept an authority which was questionable? The leading organs tried to solve these problems by breaking up the groups, by removing or even deporting individuals who seemed to be unassimilable. While the Party, as a constituted body, tried once again to become a coherent bloc, some of its leaders used its diverse elements in an attempt to play their own political game. Confronted by these men who had experienced former quarrels over the succession, who wanted once and for all to guarantee their own security and power over any potential rivals, did Stalin allow them free rein because he was ageing? Or did he aggravate the conflicts in order, once again, to defend his power and to destroy those whose ambition it was to succeed him? Whatever explanation one accepts, Stalin's position was decisive in the quarrels, and in the reckonings of the leaders which developed against the background of the great problems of the USSR; economic reconstruction and the re-establishment of a lost unity. In these conflicts two factors predominated: the antagonism between Zhdanov and Malenkov, the rise of Khrushchev at the expense of his rivals.

THE POLITICAL CONFLICTS AT THE SUMMIT

The conflict between Zhdanov and Malenkov seems to have begun during the war and was only ended by the death of Zhdanov, who died of a heart attack in 1948. The two men became increasingly influential during the pre-war years. From 1934 Malenkov occupied an important position in the Party at the head of the Directorate of the Cadres in the Secretariat of the Central Committee. In 1939 he was a full member of that Committee, of the Secretariat and a candidate for the Politburo. The war brought him new responsibilities, probably because of his organisational skills which were recognised by his adversaries as well as by his friends. In 1943, he was placed at the head of the Committee for the Reconstruction of the Liberated Areas and became, after Stalin, the most important person in the political hierarchy. In 1944 he took charge of the Committee for the Dismantling of the German Industry which was to guarantee the preparations to the USSR.

While Malenkov, after a brilliant career in the Party, advanced in the technical branches, the rise of Zhdanov took place within the apparatus. In 1934, he had succeeded Kirov as leader of the Leningrad organisation and arrived in Moscow in 1945, where he was appointed to the Party Secretariat. In the hierarchy, his place was immediately behind Stalin and Malenkov, and in the period of eclipse which the Party underwent after the war, Malenkov, who was responsible both for State tasks and responsibilities within the Party, seemed to be the heir apparent. In the quarrel which was to set the two men at loggerheads, Zhdanov relied for support on one of Malenkov's collaborators in the Committee of Reconstruction, the economist Voznesensky, author of an important book on *The Economy of the USSR during the Great Patriotic War.*

Zhdanov and Voznesensky continued the conflict at the technical level and attacked Malenkov on the policy of the dismantling of the German industries which led, they said, to waste on a vast scale. Mikoyan, who had also worked with Malenkov in 1943, was given the task of conducting inquiries on the spot and returned with a report which was very unfavourable to his former chief. He concluded that the policy of dismantling must be replaced by the setting up of mixed companies in Germany which would organise production in the service of the USSR. Stalin accepted these conclusions and relieved Malenkov of his post, at the same time removing him from the Secretariat. This, it seemed, was disgrace and the elimination of Zhdanov's only rival. But the disgrace was short-lived. After an eclipse of two years, Malenkov reappeared in the Secretariat of the Party in the summer of 1948.

Some weeks later, Zhdanov died, leaving his supporters face to face with Malenkov, who proceeded to a vast purge. Voznesensky, at the time President of the Gosplan and Vice-President of the Council, was stripped of his position and shortly afterwards liquidated. The same fate befell other leaders: Kuznetsov, Secretary of the Central Committee, Rodyonov, President of the Council of Ministers of the FSFSR, Popkov, Secretary of the Leningrad organisation. The theoretical journal of the Party, *Kommunist*, edited by Voznesensky's brother who was also eliminated, was purged for having supported the arguments of Nicholas Voznesensky. The rivalry between the two men ended at this time in a very traditional purge in which Stalin took no apparent part, simply abiding by the opinions received on the technical problems. However, given his power at the time, his neutrality cannot be taken for granted. It seems rather that he deliberately encouraged men whose personal power had become too great, to exhaust themselves in a futile conflict. The argument has been advanced that Stalin, having weakened Malenkov, turned against Zhdanov because his position was based on the Leningrad apparatus, which, because of its independence, had always been in conflict with the central authority. In support of this argument, it can be said that most of those purged had belonged in the past, or at the time of their fall actually belonged, to the Leningrad apparatus. Could this apparatus have played a decisive role in the conflicts of those in power? The documents published after 1956 are silent on this subject, but it seems that the Yugoslavs at least, when their relations with Stalin became tense in 1948, had thought of seeking support from the Leningrad organisation and Zhdanov to prepare a change in Stalinist power. It is impossible in this sphere to go beyond guesswork, but this would possibly explain the bloody character of the purge of 1949 at a time when the purges were bloodless.

Malenkov's return to power enabled him to impose his views on the Party organisation. He suppressed the Directorate of Cadres, set up by the XVIIIth Congress (in 1939) and returned to the system created in 1934 by Mikoyan. The specialised economic sections were returned to the Secretariat whose responsibility it was to control the various sectors of the economy. This abandonment of centralisation was designed, it seemed, to enable each branch to be more rationally trained; it was probably a sign of the Party's concern to supervise closely the economic evolution during the difficult period of post-war recovery.

During the same period in which, after 1948, Malenkov occupied a central place in the Party organisation and devoted himself to making it efficient, another leader began to impose himself through a series of technical positions: this was Khrushchev. In the years 1946–47 in the Ukraine, where he held both governmental and Party responsibilities, he had passed through a very difficult moment, for some months being eliminated from the leadership of the local Party. At the end of 1947, however, he regained his posts and arrived in Moscow two years later in a very strong position. His attitudes towards agriculture, even though they did not always meet with universal approval, showed that he was the most competent technician of this difficult problem, since on most of these questions he succeeded in eliminating the specialists who had hitherto directed the agricultural sector. Thus, Malenkov did not remain for long as the potential successor and the rivalry between Malenkov and Zhdanov was succeeded by a rivalry between Malenkov and Khrushchev. The XIXth Congress (1952) was to confirm that these two men stood out from among the other leaders and took up positions directly behind Stalin.

Stalinist power seemed thus reduced in the post-war years by the personal conflicts which prematurely opened the conflict over the succession. This led to new questions. Was Stalin really superseded? Were these conflicts new and important? Were they a particular factor of post-war Soviet politics? A careful examination of Stalinist policy suggests that these arguments do not hold water and point to a genuine political coherence.

After 1945, Stalin was faced with many tensions in society, which were to increase until 1952. He solved them brutally, but with due regard to the eventual consequences of his brutality. This was true, for instance, of his reaction towards the nationalities. He had foreseen that the most serious weakness of the USSR in 1941 had been that of the Ukraine. But to punish the Ukrainian people for collaborating was to run the risk of raising against the government a people who, after the Russians, were the most numerous and whose economic role was very important. By striking at the small isolated peoples, he punished the centrifugal movements by example, and warned the nations that their destiny was inseparable from that of the USSR, while at the same time avoiding dangerous consequences.

The same was true of the changes of title which he imposed on the leading organs and on the army after the war. By so doing, he ceremonially broke with the USSR of pre-war years, which had been condemned by the moral collapse of 1941 and implicitly affirmed the existence of a new State.

The coherence of Stalinist power can also be found in his attitude towards the personal conflicts and tendencies which seemed to divide the post-war Party. The debates about ideas no doubt existed, arising out of the relative liberalism of the years 1941–45, but they should not be linked too closely to the conflicts between individuals. To take one example: Voznesensky had been among the most vehement opponents of Varga. However, the fall of Voznesensky after Zhdanov's death did nothing to change the general hostility to Varga's theories. The position of individuals seemed to be linked rather to the problems of power than to differences of opinion. It seemed that Stalin, aware of the latent personal conflicts, aware also of the political impatience of some leaders and finally of the general difficulties with which his country was faced, had tolerated, even encouraged, the personal conflicts to set his colleagues the one against the other, to make them responsible for the various problems of the USSR and to assume the role of arbiter over the individuals and the tensions. This was a renewal of the Stalinist method of the 1930s which had led him to total power and which, after 1945, enabled him to maintain his absolute authority. In this perspective, the personal conflicts were dimmed by Stalin's determination to efface or mask the social contradictions and the centrifugal ideological tendencies, and in the last resort one has the impression that far from being the plaything of those who, behind his back, fought to be his successor, Stalin had never stopped playing with them.

POST-WAR STALINISM (1950–1953)

Although during the years of reconstruction Stalin tried to maintain the return to authority within an acceptable framework, after 1950 the situation deteriorated because of the purges. However, until the end of Stalin's reign, one thing remained remarkable: this was that the purges were restricted to particular categories, Stalin controlled the development of the terrorist system and at no time did it degenerate or threaten the whole country. Although the 'Stalinist paroxysm', recalled by Stalin's daughter, undoubtedly existed between 1950 and 1953, it was confined to certain groups. For the majority of the people this expression meant above all very hard living conditions, the reappearance of almost irrational methods of management and the growing repression and centralism of the government. The group upon which the repression of the years 1950–53 was to weigh most heavily was described at the time as 'cosmopolitan' and 'Zionist', and it is clear, here again, that the Stalinist philosophy was not inspired simply by blind anti-Semitism as Svetlana Alliluyeva stated, nor by pure and simple madness, but took into account the external problems confronting the USSR: its attitude towards Israel, the rapid spread of socialism in Eastern Europe and the need to consolidate the monolithic structure.

The creation of the State of Israel had created many problems for the USSR. It recognised the new State very early, both because the fate of the European Jewish community during the war forced the nations to support their demands and also because the USSR hoped that Israel, in a Middle East which was still dominated by the Western Powers, would be a magnet for social progress, for political transformation which would finally separate the whole of the region from the West. The Arab reaction, the bitterness of the Arab communists in face of what they saw as a betrayal, made the USSR aware in 1950 it had been a disastrous error to support Israel which was politically orientated towards the Western Powers. The support given in 1948 to the small Jewish State had definitely embroiled the USSR with the Arab mass. On the internal plane also, this operation had unfortunate results. Attracted by the existence of a national State, many Soviet Jews asked to be allowed to emigrate. To accept this emigration was to recognise that the national loyalties – and again Marxist theory was confused about the question of Jewish national loyalties – were stronger than social loyalties. It was also to recognise that religion could be a factor of national unity, which raised innumerable problems in the USSR itself. Stalin therefore attacked the 'cosmopolitanism' of those who wanted to go to Israel or of those who took too great an interest in it. Without yet stretching out a hand to the Arabs, he showed them that he was no longer inclined to strengthen Israel. Concern for an improvement of the Soviet position in the Arab world was a factor in this switch which was apparent at the end of 1948. In the ensuing years, arrests and deportations proliferated among those suspected of cosmopolitanism. Many of those arrested were Jews, so that the campaign against cosmopolitanism had a strongly anti-Semitic tinge.

The evolution in the popular democracies confirmed this impression. From 1949 to 1952, the popular democracies underwent two successive waves of purges, the significance of which changed each time. The first was directed against 'bourgeois nationalism', the second, as in the USSR against 'cosmopolitanism'. In the first stage (in which were engulfed temporarily, or for good, Gomulka in Poland, Rajk in Hungary, Kostov in Bulgaria, Clementi in Slovakia among many others), it seemed that the 'national' political leaders were being eliminated and replaced by Muscovites, closer through their past to the USSR. This elimination enabled the people's democracies to be linked more closely to the USSR and their leaders subordinated to those of the USSR.

However, this interpretation has its flaws. In 1947, Stalin knew that national interests could bring into opposition to the USSR men who were linked to it by the past, and by their membership of the Comintern; this was the case with Tito. At the end of 1951, another trend could be seen in the purges: this was shown by the case of Slansky. Slansky seemed to conform to all the Soviet demands: he was very close to the Soviet theses and it was he who purged all the national cadres of his country. In the trial which was to lead to the death of Slansky and his collaborators, there were eleven Jews among the fourteen accused. In this second phase of the purges, in which *cosmopolitanism* was the key word on which people were questioned and condemned, there disappeared precisely those communists who had lived for a long time in Moscow, who seemed to everyone to be the mouthpieces of the USSR, such as the Romanian Anna Pauker, and the majority of whom were Jews.

Parallel to these arrests and trials which were less explicable than those of the anti-nationalist wave in Eastern Europe, the situation in the USSR suddenly grew more tense at the beginning of 1953 and was oddly reminiscent of the situation which existed in 1936. On 13 January 1953, *Pravda* announced the arrest of nine doctors accused of crimes similar to those imputed to the accused in former times; they had murdered Zhdanov, prepared the murder of several marshals: Vasiliyevsky, Konizhev, General Shtemenko, etc. One detail in this affair was particularly amusing and tragic and conformed to the logic of the system of terror which had been seen in the 1930s. One of the nine doctors denounced on 13 January had been called as an expert in 1938 in the trial of Bukharin and had accused the whole group of the Right of having murdered Gorky and of having prepared the murder of Stalin; he himself was now accused of the crimes of which he had accused those who were on trial, and this transition from the position of accuser to that of accused was a repetition of something which had already been seen many times.

But what also characterised the 'doctors' plot' was that seven out of the nine were Jews, and as regards the accused in Eastern Europe, one fact emerged from the charges established against them: they had taken part in a Zionist plot, and had acted at the instigation of the international Jewish organisation, the Joint. No sooner was Stalin dead than in March 1953 *Pravda* published the news of the liberation and rehabilitation of the doctors who had been the victims of a 'machination'.

But before they were saved by Stalin's death, the USSR found itself plunged into an atmosphere of exceptional tension. The press published appeal after appeal for vigilance, and insisted on the idea that the progress made by the USSR 'led not to the dying down but to the heightening of the battle'. Was there to be a return to 1936? The purges in Eastern Europe and those which were looming in the USSR seemed to be part of a coherent whole. Was this the consequence of Stalin's senility which in the end assumed a bloody form, in which the anti-Semitism which he had never tried to hide completely unfurled, linked, stated his daughter, to his never extinguished hatred for his indestructible adversary Trotsky? History is full of examples of tyrants who at the end of their lives declined into complete madness or senility but even in 1953 this did not seem to be an adequate explanation for the last version of Stalinism. If one looks closely at all those who were persecuted by Stalin at that time one can just discern the causes of that persecution; one also glimpses that under the appearance of uniformity – hostility towards Zionism, even a degree of anti-Semitism – the end pursued by Stalin was not the same in the people's democracies and in the USSR.

In the popular democracies, it can be seen that the victims of the 'anti-cosmopolitan' repression had, in fact, regardless of their origins, some points in common that a book like Arthur London's *On Trial* helped to illuminate; often they had been members of the International Brigade; now Stalin had always taken care to keep out of the Party men whom the Spanish Civil War had brought into contact with anarchists and Trotskyists and who had generally been very critical of the policy of the USSR and of the Comintern. Eliminated in the USSR, many of them had led the resistance in Eastern Europe and in 1945 found themselves in the national parties and in power. Moreover, among the victims of these purges another category of communist was to be found, the Muscovites, those who had never shaken off Soviet influence and who had spent the most important part of their lives in the USSR. At first sight, their elimination was pure folly. But their case was more complex. They had lived in Moscow, in the circles of the Comintern, through a terrible period in which they had been witnesses of the purges, of the elimination of the old cadres of the Comintern, in which they had seen Stalin decimate all the apparatuses of the national parties. Their lives had hung by a thread, for no one was sure that he would escape the terror. Those whom Stalin eliminated along with the former members of the International Brigade – witnesses of his refusal to support the Spanish revolution – were the old guard of the national parties. He continued against them the operation already carried out against the Bolshevik Party, and created in Eastern Europe Stalinist parties similar to those in the USSR. The fight against Zionism seemed from this point of view rather to be a pretext, masking the creation of monolithic structure. In Eastern Europe, in spite of the existence of strong Jewish communities, anti-Semitism had deep roots among the peoples and the masses could accept more willingly purges which struck at a particular class of citizens than the determination to take from them leaders who really represented their nation.

In the USSR, one could also understand the significance of the 'doctors' plot' if, instead of linking it to the anti-cosmopolitan campaign of the preceding years, it was seen as the beginning of a more important movement, of a radical purge of the political personnel of the USSR. There are sound reasons for this suggestion. In the first place, we have Khrushchev's revelations at the XXth Congress (1956). He had, at first, stated that the doctors' plot was to give the signal for a purge of the political personnel at the highest level and quoted among the most threatened leaders: Molotov, Mikoyan and Voroshilov.

This argument is strengthened by the publishers of the successive editions of the *History of the Communist Party*. Although, in the post-Stalinist variants, the doctors are barely mentioned, the authors stress constantly the deterioration in the internal situation, and the dangers of the personality cult to the political personnel.

Curiously enough, the publishers of the *History of the Communist Party*, who generally accuse Beriya of having helped to develop the system of terror, acquit him of the increased severity of 1952–53. This would seem to imply that for once, not only had Beriya taken no part in the purge which was being prepared, but that like the other leaders he would have been one of its victims. In fact, Beriya was under definite threat during this period. The purges carried out in Georgia in 1951 affected above all the Mingrelian organisations and cadres; now Beriya was a Mingrelian and surrounded himself with his compatriots; the Mingrelian national plot, denounced in 1951, had meant that many of his friends were purged. Similarly the suggestion that there was an international Jewish plot evoked through the doctors' affair affected him directly. During the war, it was Beriya who had set up with Mikhaels the Jewish Committee of the USSR and had established the links with the international Jewish organisations. The very modalities of the way the plot was unmasked placed Beriya in a difficult situation. Dr Timashuk who had denounced his colleagues had 'had to intervene with Stalin in person for his warning to be taken seriously'. *Pravda* waxed indignant about the 'heedlessness of men in our country' which was equivalent to sabotage; was not Beriya whose police had failed to uncover such a dangerous plot implicitly described as one of those careless people?

What was the reason for this decision to purge the ruling apparatus once again? There are two explanations. First of all, Stalin, who had rebuilt his power after the war, possibly wanted to base it on new cadres. From 1930, he had continually got rid of his colleagues. Was it plausible that in 1939 he had suddenly decided never again to eliminate men who were potentially his rivals? After more than ten years of tranquillity his colleagues had acquired solid positions and probably personal support in the State apparatus or

Party. The competence and the functions exercised for so long all helped to strengthen the position of the political leaders of the USSR in 1953. Furthermore, Stalin was already seventy-three and his health was failing. It was inevitable that his colleagues should be thinking of who would succeed him; besides, he encouraged them to do so. Stalin could not tolerate these rising ambitions at his side, especially when they were supported by any kind of social base.

Another characteristic of this period must be taken into account; this was the appearance in the Party and State hierarchy of a new generation. The figures provided during the Congress of 1952 give a very clear picture of the situation in the Party, but it is clear that what was true for the Party was also true for all the ruling cadres. In 1952, the Party was in an abnormal situation. At the top and in the key posts of the secretariats were men who had lived through the period of the purges and who were for the most part over fifty. Among the rank and file because of new members, the Party had been considerably rejuvenated. In 1947 for 1 member of over 50, it contained 4 members between 40 and 50 and 8 aged between 30 and 40. At the time of the XIXth Congress, although the relation of the members aged from 30 to 40 and from 40 to 50 tended to be equal, the men who were over 50 always formed one-quarter of the upper age bracket, and for the most part they held the higher posts. Behind them, the ambitions of the younger members were frustrated and they could not advance. All the important positions were held by the 'Old Guard' or by men who had come into the Party during the war. For the new, younger recruits, the prospects of advancement were all the more restricted in that above them, apart from the summit occupied by those who had escaped the purges, were fairly young men who blocked up the hierarchy. Moreover, these recruits, above all those of the war years, had often thrown off the grip of the Stalinist administration and owed their entry into the Party to their military exploits or to channels other than the Party. To allow this generation with different outlooks and loyalties to mark time and to grow impatient implied a real danger to the Party as Stalin saw it. Finally, to these purely political motives were added technical ones. At all the levels of the State and the economy the tasks of management were complicated. The USSR was now able to respond to this because of the very real progress in the level of education. The composition of the Party reflected this progress. In 1939 the Party contained 5.1 per cent of members with higher education, 14.2 per cent with secondary

education; in 1947 the percentage of members with a higher education rose to 6.3 per cent and secondary to 20.5 per cent. In 1956 it
was to be 11.2 per cent and 25 per cent.

Thus the USSR now possessed cadres capable of assuming the
new more complex jobs, better fitted for those jobs than the cadres,
often political in origin, which had been in charge of them during
the period when the scarcity of competent personnel was particularly acute. A purge of the Party would enable the whole political
administrative and intellectual élite of the Party to be renewed.
Clearly Stalin had become aware of the inability of the ruling élite
to deal with the situation in the USSR. This was, after all, an élite
broken by the purges, material sacrifices, and by fear. To renew
the ruling apparatus once again and with it the whole administrative and economic structure of the USSR, was to build up again a
Stalinist élite, the unchanging foundation of his power. Everything
pointed to the fact that a purge was, from Stalin's point of view,
very logical. It was also clear what linked the purges in Eastern
Europe to those in the USSR at the end of Stalin's life. It was
a return to the system of power through terror which the war had
interrupted. It was the war which had called a halt to the purges in
1938; it was the war which had led, in Eastern Europe, to temporary alliances with other parties, and in the USSR with other ideas,
which after the war had brought about a certain ideological laxity.
After the period of reconstruction, Stalin could envisage a return at
last throughout the socialist world to a system based on the effective instruments which had enabled him in the past to eliminate all
the rivals who rose up on the road of power.

THE XIXTH CONGRESS

The XIXth Party Congress which was held in October 1952,
seemed to confirm that the Stalinist violence of the 1950s had a
rationality and a logic which was understandable. The Congress in
fact revealed the problems of the USSR at that time and the evolution of the Party. The Congress, which met on 5 October 1952 was
the first to be held after the war. Thirteen years separated it from
the preceding one – something which ran counter to all the rules of
the Party. This first official manifestation of the Party for such a
long time was all the more important in that during the intervening
years the Party had become weaker, almost overlooked, and that
the Congress seemed to mark its rebirth.

On the ideological level, the Congress made few innovations. It was completely dominated by the problems of organisation, made urgent because there had been no meeting for years and by the changes which had taken place in the Party.

However, ideological concerns were also present at the Congress but formed the backcloth for their debates. Stalin, shortly before, had published his thinking in *The Economic Problems of Socialism*, and to some extent this book contained the potentialities for change: it carried the seeds of the coming reorganisation. Stalin stressed the new objective of the USSR's internal policy: the transition from socialism to communism. Stalin gave indications of how this phase was to be directed. First of all, he let it to be understood that in the capitalist environment the USSR must concentrate on its own development (communism achieved in the USSR was definitely the *sine qua non* for the eventual change of Western society) and thus that there must be a new lull on the international stage. Stalin based his reasons for the respite on a considered redefinition of war at the time, stating that the probability of war between opposing camps was lessened by the aggravation of the inter-imperialist conflicts. While the imperialist Powers destroyed each other on the basis of their rivalries, the socialist world could advance towards communism. Thus Stalin carefully opened a door towards a hypothesis which his successors were to formulate very clearly: peaceful co-existence. Although he was not explicit about this eventuality, he did confirm that in his opinion the evolution of the world revolution always depended on the internal progress of the USSR.

His book thus also contained pointers to what would characterise the internal evolution of the USSR during the transition to communism. Stalin emphasised, following Marx and Engels, that in the phase of communism, production could be planned according to the needs of the individual members of society; but at the same time, he continued to identify the needs of society as defined by the ruling bodies of the State with the needs of individuals. It is clear that to him the transition from socialism to communism would be brought about by the ruling organs and would be the outcome of an administrative decision which perpetuated the structures of society and in the first place, the State.

It was in the light of these theoretical definitions that the debates of the XIXth Congress took place, which reorganised and strengthened the Party, the first of the ruling organs of the USSR. The formal changes which were then effected all had a significance which should not be underestimated. The change of denomination of the Party which became 'the Communist Party of the Soviet Union' (CPSU) was justified by the leadership as aligning it with the situation in the USSR. The 'b' which up to then had accompanied the name of the Party was a legacy from the opposition to the Mensheviks. In the USSR of 1953 the Mensheviks had long since ceased to exist, and it was appropriate to define the title of the Party in relation to its historical task of the moment: the building of communism. At the same time, this concern to make the name of the Party coincide with its prospects was a sign of the conviction of the leaders that, even at the stage of communism, the Party, the vanguard of the proletariat, would exist, identifying itself with the proletariat and speaking in its name instead of yielding its place to it. Finally, this would strengthen the indications given by Stalin in his book about the maintenance of the organs of authority in the USSR during the second phase of socialism.

Another formal change also revealed the profound tendencies of Stalinism. Hitherto in the list of a Party member's duties, the first place had been occupied by the knowledge and understanding of Marxism-Leninism. The XIXth Congress relegated this duty to fourth place, and replaced it by the duty to defend the unity of the Party, which was one of Stalin's constant concerns and excluded any opportunity for discussion within the Party.

The Congress also modified the organisation of the Party and the measures taken seemed to complete Stalinist policy. The first modification was the suppression of the Orgburo and of the Politburo which were replaced by a single body, the *Presidium*, which contained 36 members (25 full members and 11 candidates). It was a top-heavy executive body, with many members. This led to the creation of a smaller body, the *Bureau* of the Presidium. Khrushchev explained to the XXth Congress the meaning of this operation. The two suppressed bodies were dominated by the Old Guard, whom Stalin intended to remove from power. In replacing them by a larger body he submerged his old accomplices in the mass of newcomers, younger, inexperienced, and consequently more easily influenced. At the same time, the Bureau, being a new creation, could remain closed to the old leaders and thus it became a more important echelon of power than the Presidium. This transformation was reminiscent of the evolution of the Central Committee in the 1930s. Lenin's Central Committee had been a small, restricted and active body; its membership had been swollen in order to weaken the position of the old Bolsheviks and to shift power towards the new organs of which they were not members.

The Central Committee itself was modified by the Congress. It contained 232 members in 1952, that is to say, about double its membership in 1939. Its composition was a sign of the ageing of the old cadres. In fact the Central Committee of 1952 contained 60 per cent of the members of the Central Committee of 1939. Given that the membership had almost doubled, it must be admitted that almost the whole of the 1939 Central Committee had found their way to the XIXth Congress. This was evidence of the desire for stability of the Party apparatus. It clashed with the aspirations of the new generation of communists, and helped to explain Stalin's plans for a renewal brought about forcibly through the purges.

Finally, the Congress shed light on two problems: first of all, it showed that the USSR had been well governed by Stalin whose ideas underlaid everything and whose position alone remained untouched by the various reforms. On the other hand, it revealed the weakening of the Party after the war. This weakening was due to circumstances but also to the policy pursued by Stalin from 1941. Stalin had since that year continually strengthened the authority of the State, using in turn, or simultaneously, the two hierarchies, confusing their competences. In 1952, the Party was not the ruling organ of the USSR; it held its position through the person of Stalin, conjointly with the State apparatus. This weakening of the Party appeared clearly in the debates of the Congress. It was aggravated by the imbalance in the growth of the Party, its quasi-nonexistence in certain regions, especially in rural areas, whereas the State was omnipresent. The unsettled state of the Party was to influence Stalin's succession because his rivals, while continuing to fight for power, had to localise it, to define where, apart from Stalin, it was situated and to fight against the extreme personalisation of Stalinist authority.

The 1952 Congress clearly revealed many of the realities in Soviet politics: that Stalin still dominated the political stage and that he was deeply suspicious of his colleagues, whose positions he tried to undermine.

At the same time, the Congress showed that Stalin was well aware that he must adapt his power to the changes which had taken place in the USSR and outside. The social tensions of the 1950s were caused by the exacerbation of the citizens and their desire to escape from a system of shortages. This desire for material progress, a qualitative change, was included by Stalin in his programme; at the same time, he tried to confine to the material domain the aspirations of the Soviet people to whom any intellectual advance would also open political prospects. He defined the path to change: the passage to communism; and its limits; the perpetuation of the identification of power and of the masses. He sensed that the external world had changed and tried to see where this would lead. His last book and the last months of his life were, as were the last writings of Lenin, a genuine testament. What had to be changed and what had to be maintained were all inscribed in the actions and the texts. Stalin had seen that change was necessary, but it seems that he himself was not able to change. His death on 5 March 1953 enabled his country to take a new path: post-Stalinism.

Chapter nine

THE END OF STALINISM

In March 1953 the whole USSR saw the signs giving warning of a purge and prepared to face the return of the terror. Yet Stalin's death which took place in this atmosphere of intense collective anxiety was not received with relief. It was greeted at first with anxiety and despair. Stalin's colleagues were uneasy, fearing the political vacuum, even though it had saved some of them from an impending disgrace. But all thought back to the time of Lenin's succession; everyone looked fearfully at the others, trying to discover among their own ranks the future Stalin. They were also uneasy about the Soviet people and rightly so – this people whom they dominated from the height of their privileges and whom they did not know. How would the people react? To what extremes might they not go? The old fear of the people's spontaneity and of its sporadic revolts reawoke and united, in a brief alliance, the potential heirs around Stalin's death-bed. It was this fear that caused them to delay the announcement of the illness and death of the tyrant, and led them to surround the Kremlin with a shield of troops. A useless precaution because it was not fury, but despair, either felt or feigned, which moved the crowds which no one could control and left the streets strewn with several hundred crushed bodies. In all the work sites 'mourning meetings' were held in which the participants gave vent to their despair. Yet behind this collective despair which was caused because the people had become accustomed to the Soviet system, could be glimpsed more mixed feelings. Fear of the future. Might it not be even worse? But also the vague hope that Stalin's going could change everything; that a terrible epoch had come to an end.

Although the people's feelings about this important event – the death of the man who had ruled with an unparalleled power the life of every one of his compatriots – remained mixed, Stalin's colleagues after the first moments of helplessness, took stock of the problems confronting them. First of all Stalin's succession must be ensured; they must also decide on the decisive problem of the direction the USSR must take. Would it be possible to pursue Stalin's policy without him? Or did his death mean that there had to be new developments and changes?

THE SUCCESSION

Stalin's potential successors had a clear advantage over their predecessors who, when Lenin died, had faced the same problem. They could refer to Lenin's succession and draw conclusions from it. They knew the danger in allowing one man to seize power. To guard against this, they rediscovered the virtues of *collective leadership* advocated by Lenin. But, like their predecessors, they remained ill-prepared to solve the problem, because they still did not know how to define and situate the centre of power in the USSR. Two men had embodied power in the thirty-five years of the Soviet regime. Lenin, who had founded the Bolshevik Party and who after the Revolution had been only head of the government; Stalin, General Secretary of the Party at Lenin's death and who had additional responsibilities in the government. What was one succeeding in the USSR? A man? Or a function? Two constitutions, those of 1924 and 1935, had not given the answer to this problem and had not indicated how power should be handed over. Because of the silence of the texts, and of the difficulty in acting, everything had to be improvised in 1953 as in 1924.

With Stalin dead, who would emerge from the ruling group? First of all, Malenkov, who at fifty years old, seemed to all intents to be the heir apparent. He had been the *rapporteur* of the Party at the XIXth Congress in 1952; he was the main speaker at Stalin's funeral. An apparachik of the Party, he had also been trained as an engineer and because of his education had the edge on many of his colleagues. Within the apparatus, his main rival seemed to be Khrushchev, also a Party secretary. Nearing sixty, Khrushchev, who had been trained on the job in the Party schools, did not, however, seem capable of opposing him effectively. Because of his peasant habits and his specialisation in agricultural problems, he was regarded by everyone as a man of the soil, and not as a manipulator. In this over-hasty assessment his colleagues forgot to credit Khrushchev with two trump cards. First of all, his military past. He had taken part in the battles of Stalingrad and Kursk. To this he owed his links with the army which he knew how to use when the time came. He had been First Secretary of the Ukrainian Communist Party from 1938 to 1949; then First Secretary of the Obkom of Moscow in 1949. At the XXth Congress one-quarter of the delegates with the right to vote came from these two organisations. It is not open to doubt that Khrushchev found in them a group of supporters. The same could not be said of Molotov, Minister of Foreign Affairs. Although his long-standing service in the Party – he was one of the few companions of Lenin to have survived all the purges – did give him some prestige, he was a functionary rather than a political leader. The real rival was in the end Beriya, the man of the security service who ruled alone the world of camps and prisons and had at his disposal the armed forces. Of all the possible heirs, it was he who possessed the greatest real power and because of this his colleagues suspected that he had drawn up a precise plan to seize power. This is why, in the hours immediately after Stalin's death, a coalition was created against Beriya around Malenkov. He was hastily invested with Stalin's functions in the government and thus on 6 March he held, like Stalin, both the reins of the Party and of the State. With Beriya neutralised for the time being, the successors then stripped Malenkov of his excessive power by forcing him on 14 March to give up the Secretariat of the Party in order to devote himself to the government. In less than one week Stalin's heirs seemed to have found an answer to the difficult question of the succession. They had weakened the men who seemed to them most eager for power or the most likely to seize it, and set up a collective leadership in which the apparatuses were equally balanced. Malenkov dominated the State, Khrushchev the Party, Beriya the security, Molotov the administration, Bulganin the army. Stalin's centralised power had fragmented in many directions.

The appointment of Malenkov as Stalin's successor, and then collective leadership, were emergency decisions which had solved nothing. Behind the scenes, the rivalries became more acute and the struggle for power continued. But for some months it was still dominated by the anxiety with which Beriya, in spite of the opposition to his ambitions, inspired his colleagues. The instruments of his power remained and he made use of them in an unexpected way. Master of the camps, he made himself the spokesman for a policy of legality, of amnesty, which he alone was able to apply. His colleagues feared as much the concessions which he allowed society to glimpse and which would enable him to appear as a liberal, as they feared his power. This fear was enough once again to unite the leaders in a conspiracy in which, with the help of the army – necessary because of the means of defence at Beriya's disposal – they succeeded in getting rid of him once and for all. Beriya was officially arrested in June 1953, tried and executed at the end of the year. It is probable that his colleagues killed him, gave the deed legality by accusing him simultaneously of being in the service of foreign Powers, of weakening agriculture by demagogic concessions and of being responsible for the disorders which were taking place at the time in East Berlin.

Beriya's fall brought with it the fall of his colleagues in the central and regional police organisations. The end of 1953 was filled with the executions of those who had been loyal to Beriya. Was this the return to the blood-stained practices of Stalinism? To a procedure of taking power in which one must either win or be killed? No. The death of Beriya and of his collaborators marked, on the contrary, the end of the blood-stained period in the political conflicts in the USSR. After that, although the conflicts continued, although there were eliminations, never again was an adversary physically liquidated. The months that followed Stalin's death had imperceptibly begun to change the Soviet political system. In three ways. Stalin's successors were agreed that the violence of their conflicts must be reduced and that they must not get caught up in the web of liquidation, the dangers of which they knew. They also agreed to reduce the instruments of arbitrariness; Beriya's death meant that the security system could be reformed by placing it under the control of the Party and of the State. Lastly, to destroy Beriya they called on the help of the army. But there again an unwritten law seems to have imposed itself on all those who took part in the power struggle. The appeal to the army, its introduction into the political conflicts, was to be an exception. The army must be at the service of the government and not become an active element in it by acting as an arbitrator in its conflicts.

This is what separated the years which followed Lenin's death from the years which followed that of Stalin. The conflicts between Lenin's successors had been marked by a continuous growth of police power and by its interference in the problems of the succession. They had also been marked by a growing verbal and physical violence and by the rapid disappearance of every rule in the political game. In 1953, with Stalin dead and Beriya, symbol of the force of the police, liquidated, the potential successors, whose rivalries had in no way abated, were to try to find some rules limiting the range and consequences of the conflict over the succession. This regularisation of the political conflict was very evident in the period which followed. From the end of 1953 to February 1955 the struggle for the succession was circumscribed little by little to two men and to two apparatuses, Malenkov and the State, Khrushchev and the Party. It ended in Malenkov's defeat. Malenkov resigned from the government on 8 February 1955, on the grounds of his inexperience and his inability to deal with the problems of the USSR. From then on Khrushchev's ascent to supreme power began. Power shared from February 1955 to 1958 with Bulganin, who was to be at his side a colourless head of government; unshared power from 1958 to 1964. However, none of those whom Khrushchev swept out of his way in this march to absolute power was put to death. Malenkov, who had resigned, was to be criticised for his mistakes but never insulted nor expelled completely from the Party nor liquidated. He was able, on the contrary, to try once again within the Party to reduce Khrushchev's influence before devoting himself to technical activities. His deposition, which was imposed on him by his resignation clearly marked the appearance of more regular and peaceful political procedures. It showed – and Khrushchev's deposition in 1964 was to confirm this – that the Party had returned to the practices of the Leninist epoch. No matter how widely they differed, nor how great their ambitions were, the leaders no longer killed each other. And although to give up power meant certain political death, this political death did not lead to physical death.

Nor was the 'dedramatisation' of the personal conflicts the only change in the Soviet system in the years between 1953 and 1956. A new political practice must be mentioned which was characterised by a more regular functioning of the institutions. Until March 1953, all the institutions, those of the Party and those of the State, functioned outside the rules, at the bidding of Stalin's whims. After his death, his successors tried to revive apparatuses which had been reduced more or less to ghosts. The Presidium of the Central Committee was to meet nearly every week; the Central Committee was to be convoked six times in the three years which separated Stalin's death from the XXth Congress. It is clear that, in these conditions, the personal conflicts became known outside the small group of the Presidium and that the Central Committee could intervene in them. The State, too, returned to its normal practices. The Supreme Soviet which, under Stalin, had met only once a year, after 1957 met twice a year. Above all, in the governmental bodies decisions were taken and work carried out during normal hours and no longer, as in Stalin's day, during interminable nights of heavy drinking in the dachas where he summoned his horrified and frightened collaborators.

The years 1953 to 1955 were thus decisive for the Soviet political system because they introduced the implicit will to institutionalisation. They were also decisive because the conflicts over the succession were played out against the backcloth of a fundamental political choice: how to preserve the Stalinist orientations?

THE CHOICES IN THE BREAK WITH STALINISM

From 1953 to 1956, the Soviet system remained officially unchanged. No public debate indicated that Stalin's decisions were to be challenged. Nevertheless, in three fields, the practice and options of Stalin's successors clearly separated themselves in that time from Stalinism: in the economy, in policy and in international relations.

In the economy, Stalinism had been characterised by the priority constantly given to the industrialisation of the USSR, to the detriment of the needs of the citizens. On this plane, the change took place very quickly, even though the results were not always convincing. Malenkov proclaimed in 1953 that the time had come to consider the needs of the people and of the Soviet consumer. To achieve this, he lowered prices in 1953 and 1954, then launched a campaign designed to develop the production of consumer goods, and revised the Five-Year Plan. The emphasis given to the improvement of the standard of living came not only from these overdue measures but from a change in the political vocabulary. The slogans which greeted the anniversary of the Revolution in 1953 concentrated on the right of the consumer, on the search for wellbeing, on the transformation in the standard of living. The emphasis was no longer laid on heavy industry and equipment, on the global interests of the USSR; it was stressed that the destiny of individuals was also equally important. If one adds to this the fact that the Supreme Soviet opted in October 1953 for tolerance towards the peasants who had not fulfilled their production quota, one can see that a genuine economic revolution was taking shape.

Of course, in the first stage, this revolution was disappointing. The fall in prices which was not accompanied by a growth in the products to be found in the market brought moments of great shortages. Malenkov's agricultural programme ran into technical difficulties (insufficient fertilisers and machines) and clashed with the demands of the programme of clearing for cultivation the virgin lands sponsored by Khrushchev. Generally, the rivalry between the two men – Khrushchev by his ambitious agricultural projects sabotaged the more immediate policy of Malenkov – was a serious brake on economic progress. But, in 1955 when Malenkov fell, Khrushchev assumed responsibility for the initiative of economic change. He advocated that in agriculture an increasing role must be given to the initiative and industrial activity of the kolkhozes. In industry, he tried to give the technicians a more important role, to those who were trained, at the expense of political controls; and the idea of a degree of autonomy in the enterprises began to be debated. Undoubtedly, unlike Malenkov, Khrushchev did not go so far as to want to suppress the old Stalinist priority of heavy industry. In giving priority to the industry of consumer goods, Malenkov had clashed with all those – the armed forces and the bureaucrats of heavy industry – who had hitherto had a privileged position. Prudently avowing himself loyal to the traditional economic trends of the USSR, Khrushchev was content to place the two industrial sectors on an equal footing: and proclaimed that what was important was efficiency and not conflicts between schools of thought. Having done this, he gradually improved the economy and won the loyalty of a society which for the first time since 1930 saw that its needs were being recognised. In 1955, an economic revolution took place in the USSR. The peasant realised that the decrease in the continual administrative harassment he had been subjected to had died down and that he was no longer treated as an enemy. Even more, everything suggested that the era of continual revolutions which was to lead to a complete suppression of the kolkhoz was over. Although Khrushchev had not yet spoken of the ultimate direction society was to take, his concessions and the insistence on peasant initiative suggested that the stabilisation of the countryside was included in his programme.

The political conditions of the citizens also changed at that time. Limited measures of amnesty were adopted in 1953. In September 1955 an amnesty for the crime of collaboration was proclaimed. It was not applied to all those who filled the camps as collaborators (how many prisoners were there accused of this crime?) but nevertheless it brought some measures of liberation. Detainees began to leave the camps to make Soviet society aware of the world of the concentration camps, to ask the question as to whether it should continue. Above all, the problem of arbitrariness, of the purges

and internments gradually emerged. The government dealt carefully with this problem because they were aware of the risks inherent
in a radical change. The first measure which challenged Stalinist
despotism was passed immediately after Stalin's death. On 4 April
1953, a communiqué from the Ministry of the Interior informed
the dumbfounded citizens that the whole affair of the 'white shirts'
had been simply a huge fabrication and that innocent people had
been illegally arrested and persecuted. From that moment, the
arbitrariness of the government and the reign of illegality began to
be progressively denounced. Cases of 'arbitrary repression' recognised by the authorities were to increase until 1956.

In September 1955 a trial took place in Georgia in which officials
in the police were accused of having forged the pre-war purges
from start to finish. After that, it was clear that what had been
possible in Georgia had been possible elsewhere and that the whole
system must be examined. The Stalinist world slowly collapsed.
Were those who had been treated as criminals perhaps victims?
The liberated detainees, even though the authorities demanded that
they remained silent, contributed towards this realisation of illegality and of Stalinist arbitrariness. Nothing in Stalinism was as yet
repudiated, nothing was as yet apparent, and yet everything was
ready for its rejection. The intellectual life of the USSR reflected
these changes and showed that a certain liberalisation was making
headway. *The Thaw* by Iliya Ehrenburg appeared in 1954; Dos-
toyevsky, until then a writer who was anathema in the USSR, returned to the pantheon of Russian geniuses and his complete
works could at last be published. The poet Esenin who had committed suicide in 1925 in despair at what was happening in his
country, whose name was no longer pronounced, was also readmit-
ted into the literary heritage. This was a significant return implying
as it did a disavowal of Stalinist dogmatism. This dogmatism began
to be discussed in the pages of historical journals and the toll it had
taken in the intellectual and scientific progress of the country began
to be assessed

International relations were another field of innovation for the
successors. Stalin had based his post-war foreign policy on two
principles: absolute opposition to the external world, characterised
by the Cold War and the complete domination of the other socialist
countries. After 1953 Soviet foreign policy became more supple.
The end of the Korean war, the end of the war in Indo-China, the
treaty signed with Austria, the overture to the Third World to
which Khrushchev did not preach fratricidal revolutions but only a

consolidated independence and national development, were all signs of a new approach to the outside world and of a determination to become part of it instead of purely and simply opposing it.

But it was in relations with Yugoslavia that Stalin's successors were to show just how far they had put the past behind them. Stalin, when he had realised that he could not manipulate Tito, and could not impose his will, had broken with Yugoslavia, had expelled it from the socialist community and had tried, by brutally suppressing any aid, to bring it to its knees, but without success. His successors, by opening the USSR to the world outside, had realised that such a policy must be accompanied by a change in relations within the socialist world. Relations must be more fraternal than violent in order to find the basis for a new loyalty. The reconciliation with Yugoslavia was a necessary element in this cohesion. But here, the agreement was far from being complete. Although Khrushchev was by all accounts the promoter of such a reconciliation, he clashed with Molotov who defended in *Pravda* the Stalinist position in the conflict with Tito. The Yugoslav affair contained all the stages which enabled the changes in the Soviet political system to be measured. The debate between Molotov and Khrushchev on the attitude to be adopted towards Yugoslavia was opened in February 1955. On 9 March, *Pravda*, the Party's official newspaper it must be remembered, published in response to Molotov a speech given by Tito two days earlier which set out the Yugoslav thesis and put Stalin and the Soviet policy in the wrong. That this position could be expressed in the Soviet press, that the debate should be public was indeed something new. Without doubt, the debate hid the conflict between the two leaders of the USSR, who were still fighting for power. But because of their struggle an important debate took place in public on a subject of the greatest importance. Had Stalin been right or wrong in his relations with Tito? It was Khrushchev who brought back the answer after his journey to Belgrade in May 1955. Having gone to Yugoslavia to make peace with Tito, Khrushchev had, in the last resort, disavowed Stalin's action in 1948 and even more the principles which until 1955 had guided USSR's relations with the popular democracies. He not only recognised Stalin's mistakes, but accepted a final declaration which declared that the relations between communist States were relations of State in which the solidarity of the parties was not involved. In other words, he recognised that the nature of relations between communist States was not specific in any way: that a communist State was an independent State, similar to other States, without any specific links with States of the same system, and that each communist party alone made its choices. He recognised that there was more than one road to communism. National communism and the end of communist internationalism were the concessions made by Khrushchev to Tito in May 1955. This represented a complete break with Stalinism. Stalinist ideology and the practices which had been linked to it had to be completely revised.

The years which separated Stalin's death from the XXth Congress can be seen as years of erosion: as years in which those in power tried to respond to the needs of individuals and to some extent to take into account, not only their material but also their political aspirations. All Stalin's successors were agreed on the need for changes in the USSR. But they differed about the methods to be used and the timing. Men like Molotov and Kaganovich feared above all that the changes would get out of hand and that they would find themselves acting as the spokesmen of a society greedy for more, and not as the trailblazers of this movement. The consensus of the political class that a transformation in the system of government in the USSR was essential ran parallel with the agreement on the limits which must be set to it. For all the leaders the system had to be changed in order to take into account the changed internal and international situation, but the essential element must also be preserved, the Party's monopoly of power.

THE LIMITS OF EROSION: A STRENGTHENED PARTY

During the two years in which he was overshadowed by Malenkov, Khrushchev was to influence decisively the future of the Soviet political system by restoring to the Party the central role of which Stalin had deprived it. In 1953, the Party had been weakened by the successive purges which it had undergone, by the demoralisation of its cadres and by the importance given by Stalin to the State. Khrushchev who, in 1953, was only one secretary among others and the fifth in the Presidium in order to precedence, was to devote himself to two tasks: to reconstruct the Party and to ensure its pre-eminence. In this respect, his activity in the years 1953–55 recalled Stalin's activity after 1920; and the attitude of his colleagues towards him was similar to that of Stalin's colleagues. Although Stalin had been able, in the 1920s, to acquire considerable power, this was both because his colleagues had a low opinion of him and because they did not understand the importance of the apparatus he was creating. In 1953, Khrushchev, by his peasant ways, his loquaciousness, the vigour of his language and his crude behaviour upset Soviet ideas of behaviour. He was *nekulturni* (uncultured) in the eyes of his colleagues, unrepresentative of the high-ranking Soviet bureaucracy. And this meant that he was not seen as a genuine rival. This is why his colleagues allowed him to take over the administrative jobs of the Party without realising how he was rapidly rebuilding this instrument of power.

Khrushchev's power was to grow all the more quickly in that it was intermingled with that of the Party and that Khrushchev's demands were formulated in the name of ideological strictness. The take-off for this double rise was the Plenum of the Central Committee which took place between 3 and 7 September 1953. A month earlier the head of the government, Malenkov, had announced, without consulting the Party, the economic reforms which he was putting into force. By so doing, Malenkov, as Stalin had so often done before him, gave the primacy to the State apparatus and asserted its political autonomy. The September Plenum gave Khrushchev the opportunity to redress the situation. It was he who convoked it. The agenda included, at his instigation, the agricultural problem and he used the occasion to recommend Malenkov's entire economic programme, presenting it as the initiative of the Party. At the same time, and for the first time, he revealed the weaknesses and failures of the Soviet economy, and concluded that in order to succeed, a rural programme for the countryside needed the collaboration of the Party organisations. In doing so, Khrushchev reached three goals simultaneously. He gave the Party a share in governmental initiative which was very popular as it responded to the aspirations of society. Further, it showed that the Party was the dominant body in the USSR; that the decisions taken within the State apparatus could have no validity unless they were taken in agreement with the Party and ratified by it. Finally, at local level, he brought the reforms under the Party's control and despatched many communists to the countryside to strengthen the supervisory organisations. By so doing, he inserted the Party into all levels of the economy, thus strengthening its role.

Because the Party was to assume wide responsibilities in the economy, the man who was its real leader had to be given a status befitting his responsibilities. The Plenum during which the rebuilding of the Party was begun approved Khrushchev's position by appointing him as First Secretary.

Beriya's fall in 1953 had given Khrushchev a further opportunity to act within the Party. In ferreting out Beriya's accomplices,

Khrushchev purged the Party organisations in Leningrad and the Caucasus. These measures, which were popular because they were designed to punish the agents of the repressions, enabled Khrushchev to place new men everywhere who formed the basis of a party loyal to himself.

In February 1954, he pushed his advantage still further. During the February Plenum he proposed 'his own' plan to use the virgin lands to solve the agricultural problems of the USSR. He then took the initiative in the economy, without consulting the government, and showed once and for all that the true decision-making centre was the Party. Because his programme pleased those concerned with heavy industry because it stated that heavy industry should not be sacrificed to consumer needs, because in the military debates he argued against Malenkov that the existence of nuclear weapons did not mean that there was no danger of war, thus preventing any reduction in the armaments policy, Khrushchev gained the support of the army and of industry. The Party in this way became the guarantor of the traditional options but also of efficiency, while the economic stagnation gradually discredited Malenkov.

By 1955, Khrushchev had attained his two immediate objectives. He held the highest position in the Party. And the Party once again become the main apparatus in the Soviet political system. Certainly at the time the policy over which he presided seemed to be inspired by Stalinism, even if the Stalinism had been watered down. The attachment to heavy industry, to the domination of the Party over the State (even though from 1934 Stalin had weakened the Party), and the manipulation of the apparatuses, were these not all Stalinist? But foreign policy, in which Khrushchev very soon allocated to himself an increasing role, showed that he was no more Stalinist than was his adversary Malenkov. The Party was for him the instrument of power. The struggle for the succession in which Malenkov put himself forward as an innovator forced him, as long as Malenkov was there, to defend the traditional economic arguments. But once Malenkov had disappeared, Khrushchev took over the economic reforms and the questioning of Stalin's despotism. With the Party reorganised and strengthened as the cornerstone of political life, he thought that he could forge ahead with change without by so doing destroying the system with which he identified himself. What remained, however, was that in loosening, however carefully, the constraints, in putting forward reforms, Stalin's successors constantly, ran into a problem which they had in the end to solve, that of Stalin himself. It was clearly impossible to abandon the elements of Stalinism, without reconsidering Stalin's legacy as a whole. The blow struck at him by Khrushchev at the XXth Congress was a decisive stage in the process of the liquidation of Stalinism and certainly constituted Khrushchev's greatest merit in the eyes of his contemporaries.

THE XXTH CONGRESS: THE CONTROLLED TRUTH

When the XXth Congress met, the USSR had already changed profoundly. The many reforms since 1953, the new attitude towards the outside world and the still restricted opening of the camps had created a new political climate which was sensed throughout the whole society. At the same time, an underlying anxiety continued to exist: would the change last? Or would it lead to a new period of tyranny? Stalin had accustomed his compatriots to these frightening changes, to terrible hardenings of attitudes. The fluctuations in the attitude of the Soviet leaders towards Stalin could be used as the justification for every fear and every hope.

It was a sign of the difficulty they found in knowing how far they should go in the changes they had initiated. With Stalin dead, his successors had at once tried to rid the political stage of his presence, by keeping silent about his name which until then had filled every public demonstration. In 1953, the disappearance of Stalin's name from the newspapers was a striking sign of their determination to eradicate the dead leader from Soviet history. Only Lenin, hitherto associated with Stalin, was mentioned. However, at the end of 1954, the memory of Stalin reappeared. His birthday, passed over in silence the previous year, was celebrated with some pomp as shown by *Pravda* of 21 December. The following year, on the same date, the celebration was colder, but Stalin's photograph in *Pravda* showed that he still held his place in the history of his country. Some days later, however, the needle swung into the zone of silence. At the beginning of January 1956 the USSR – in which the custom of keeping anniversaries is highly developed – celebrated Voroshilov's seventieth birthday. Tradition demanded that the official good wishes sent on this occasion and published in the press be accompanied by the formula 'companion of Lenin and Stalin'. Then Stalin slid into oblivion.

The changes in the attitude to be adopted towards Stalin were the expression of the leaders' doubts and probably of genuine disagreements. But while the official attitude towards Stalin was changing from day to day, a process of rejection had begun in 1953 which developed less visibly in the background and which continued to grow. The Party had set up a working group for which Pospelov was responsible; this group was to assemble material which would enable Stalin's achievement to be assessed. It gathered its own information and used the material assembled by the commission of rehabilitation whose discreet work nevertheless remained unchallengeable; it also benefited from the research in which the Soviet intellectual élite was engaged. It was in this way that the *Historical Institute of the Academy of Sciences* under new leadership examined what had taken place in the USSR in the troubled years since Lenin's death, Stalin's own actions and the way in which history had been written. It called a great conference, no longer – and this was a profound revolution of which everyone was aware – to deliver devastating criticisms of historical works which were not close enough to the demands of those in power, but to reflect upon the excesses of those in power. Everywhere the rethinking and the materials gathered shed light on the years of the collective Soviet tragedy. Undoubtedly those who were associated with this rethinking formed the élite and the problem was to decide whether the entire society had the right to know the truth. And how much of the truth?

The XXth Congress gave complex answers to these questions; but in spite of this complexity, it completely destroyed the remnants of Stalinism starting with the language which continued to dissimulate.

The XXth Congress had two sides. One, open to all, was characterised by the revision of a great many official decisions of Stalinism. The Congress attacked the way the Party had evolved, it denounced its caste spirit, its privileges, its bureaucratism, its lack of vitality and initiative. It also denounced the miserable condition of the workers, inevitable in a period of shortage but which the success of socialism no longer justified. It challenged the domination suffered by the nationalities in the USSR and the Stalinist conception of inequality embodied in the privileged role allocated to the Russian people. It even dreamt of restoring to the trades unions an autonomous role in the service of labour and no longer of production. In external affairs, the members of the congress in the same way abandoned the theory of the besieged citadel for an optimistic vision of international relations marked by the success of socialism and the need for nations with different systems to co-exist peacefully. The rejection of violence and the acceptance of diversity were found in the relations between socialist countries. The Soviet leaders recognised their right not to follow the Soviet model, to make the revolution and to build socialism according to their own ideas and historical past. The ideology which appeared in broad lines at the XXth Congress was in every sphere a total break with Stalin's action. It was also the reflection of the path taken by his successors for the last three years. What characterised it was the rejection of radicalism, and the desire to reconcile theory and practice. Undoubtedly, in this revision which carried within it promises of internal and external peace, Stalin's successors did not yet reply to the ultimate question which he had bequeathed to them. Should the Soviet Union take the final steps towards communism by unifying the two sectors of activity maintained in the countryside and by suppressing the last vestiges of the free market? Although the XXth Congress did not as yet abandon this remote project for bringing the social revolution to fulfilment, it nevertheless clearly stated that violence, and decisions not based on social consensus, were things of the past. The final change when it came, would this time be made with the society. Moreover, the revolutionary future was lost sight of in the revisionism which invaded all the sectors of human activity. But the fundamental innovation of this congress was to be found elsewhere; the destruction of Stalin which was taking place openly throughout the congress, but also more secretly in Khrushchev's report.

To take the overt aspect first: from the opening of the congress there were striking signs of destalinisation. It is these signs which must be followed in order to understand the congress. In fact, within the closed world of communism dominated by silence and the unique truth set forth by the Party, anything that escapes from it or is extraneous to the official truth can only be expressed by signs which must be detected and deciphered. The silence surrounding Stalin's name on the eve of the congress was one such sign. Another sign of destalinisation was to be found in the way in which the memory of Stalin was greeted at the XXth Congress, which was the first gathering of communists from the whole world since the death of the man who for more than a quarter of a century had completely dominated them. Khrushchev, and this was the only reference to Stalin he was to make, asked the delegates present to salute the memory of their comrades who had died since the XIXth Congress, such as Stalin and Gottwald. Stalin quoted randomly among others and on the same level was something that had never been heard in a communist gathering. Nor could what happened later ever have been imagined; Mikoyan attacked the personality cult, its ill-effects and particularly the cult of Stalin. If one adds that at the opening of the congress a voluminous file had been handed out which contained information and documents about the Stalinist years, especially about the purges, and which contained 'Lenin's note to the congress', inappropriately called his will, and the existence of which had always been denied by the CPSU, one realises that Stalin's trial was open and open to all and that it was no longer possible to reverse.

On the night of 15 February, a master stroke was added to the secret report. This time the delegates who were not parties to institutionalised communism were excluded from the session. The secret report was a 'family affair', reserved to those who had directly experienced Stalinism and suffered from it. Read in an atmosphere of dramatic tension by Khrushchev, this interminable report summed up Stalin's crimes, the price paid for them by his country and by the neighbouring communist parties. The report, the gist of which the foreign delegates were also to gather, but outside the all-night session, was never intended to be a complete secret. The day after the congress, the delegates were given the task of gathering around them groups of communists and even some carefully selected non-Party men and reading the report to them and discussing it with them. In the months after the XXth Congress, the secret report, which was not published in the USSR, circulated throughout the country, giving rise to heated arguments, in which everyone added his own recollections to the document, his own assessment of the system, in which the supporters of destalinisation and the appalled Stalinists clashed violently. In a few months, the contents of the secret report had become known to everyone, because those with no direct links with the Party were told of it by those who had taken part in the debates. The USSR experienced a genuine revolution of thought which a Soviet historian described as emancipation, in the sense in which the serfs were emancipated.

This expression sheds light upon the problems which destalinisation set to those in power. Why had they taken this path? How and how far were they going to follow it? In the changes and the rearrangements of the political system which the exhaustion of society and the difficulties of the succession imposed in 1953, it was inevitable that the Stalinist past should come under scrutiny. The situation demanded this re-examination and everyone undertook it for himself. Since March 1953, a dynamism of destalinisation had been unleashed because the Party and its leaders wanted to ensure their own survival. It had begun with the recognition that the conspiracy of the 'white shirts' in April 1953 had been faked, an admission which was made in order to weaken the security forces. But once a flaw in a closely knit system had been admitted, all the other flaws appeared. In this dynamism of change which was imposed on them and in which society played its part, Stalin's successors had wanted to control events. The XXth Congress was an attempt to make the point of destalinisation and to fix its limits. By attacking the problem of Stalin's personal responsibility head on, the leaders of the CP had signalled their intention to control destalinisation. Discussed, clarified, encouraged, it became the initiative of the Party, its own affair and not that of the ordinary citizens. There was clearly agreement among the political class on the essential point: the need to preserve for the Party the monopoly of political change.

This agreement took the form of the very rushed preparation of the secret report which cannot be attributed solely to Khrushchev's initiative. A report of this size – more than 20,000 words – based on innumerable details, needed profound research into the Soviet history of the preceding decades. Everything, the documents distributed, and Mikoyan's public report, even in the way the congress developed, pointed to the desire to open the debate on this point. Everything also pointed to the determination to control the debate, especially the secret report and its contents. Why a secret report when the personality cult was being publicly denounced? Probably because the Party leadership was uneasy about the consequences of such an offensive against Stalin and feared that there would be dissension when it came to deciding where to call a halt. For the most prudent – like Molotov and Kaganovich – it was better to stick to generalities and to mention the personality cult without going into detail. According to them, the congress was a test of the ability of the communist world to accept destalinisation and to enable it to go ahead without clashes. For the more radical, and here Khrushchev played a decisive part, it was necessary to go further, to trace precisely and publicly the dividing line between what had been positive in the history of the USSR and what the Party rejected once and for all. The compromise arrived at between the two points of view was that the most damaging details should be reserved for the Party and therefore that the detailed report about Stalin's crimes should remain secret. If Khrushchev read it, it was because he was anxious that the report be published. It has sometimes been suggested that by allowing Khrushchev to read the secret report, his colleagues hoped that he would be destroyed by this ordeal which the Party would not be able to accept. This is to forget that from 14 to 25 February the Soviet communists had watched, day after day, the gradual crumbling of Stalin's myth and had heard his decisions being condemned. Once the astonishment had passed, the attitude of the Soviet delegates became more and more open to the idea of the destruction of the former idol. Thus, everything was conducive to the reading of the secret report and when the long night of truth began, the Party already knew what they were about to hear and were prepared for it.

The secret report proceeded to great lengths in the denunciation of Stalin, but also contained many strange and many explicable gaps.

It went far because it said that Stalin had established a reign of terror in the USSR; that his power was founded on despotism and terror; that he had liquidated innumerable innocent people and distorted the authority with which he was invested in order to guarantee his own power. It also went very far because, denouncing the mechanics of terror, the report prevented any return later to the same mechanics. Finally it went very far because it showed that everything that Stalin had proclaimed as the truth was a lie, thus that the *truth* of power can conceal the lie.

But the secret report and the unity built up by the XXth Congress also had their limits. The secret report was characterised by three main traits. First of all, it traced a chronological demarcation in the history of the Stalinist USSR between what should be retained and what should be rejected. This demarcation was 1934, the moment when, according to the secret report, the personality cult tilted an acceptable policy towards terror and the unacceptable. Secondly, it was selective about the victims of the personality cult. According to the secret report, the Stalinist terror had expended itself on the Party and the army. The report was silent about Lenin's companions who were liquidated by Stalin, Trotsky and Bukharin for instance, and who had opposed him before 1934. Above all, and this was the essential, it was silent about the millions of ordinary citizens who had been the victims of the repression. Lastly, the secret report established a clear division between Stalin and the Party. It appeared from this conception that until 1934 the Party and Stalin had worked together to change the USSR; in 1934, Stalin had stripped the Party of all its power, had terrorised it and had involved his country in a mistaken and fatal path. These silences and demarcations are not accidental. They showed a coherent wish to control destalinisation, which radically changed the political system. In taking 1934 as the beginning of

Stalinism, the Party ratified the switch of 1929 and the methods used to apply it. It implicitly affirmed the justice of the 'revolution from above', the logical corollary of which was that the Party had the right to substitute itself for society in order to organise its destiny. In affirming that Stalin had decimated the Party and the army, and in omitting from this admission the peasants and the simple citizens swept into the whirlpool of Stalinism, Khrushchev and his companions legitimated the Party's claim to be solely in charge of destalinisation and excluded society from the process. The political system must remain embedded in its foundations, in the ideological and decisional monopoly of the Party.

Furthermore, in affirming that the Party alone, or nearly so, had suffered from Stalinism, the secret report tried to evade the problem of its complicity with Stalin because this was the heart of a fundamental question which his successors had tried to sidestep. Was Stalinism simply the product of one man's perversity? Or did it condemn the whole system? How could the Party which had cloaked, with its authority, the Stalinist lie, which had erred over and over again, be infallible? If it had erred on some points – terror, the attitude towards some foreign communist parties, such as the Polish party destroyed by Stalin in 1938 – what guarantee was there that it had not erred in all its decisions, that it had been right in 1929, that its options of change should be maintained? Finally, an accessory question in relation to the whole debate but essential for the political class: who were Stalin's accomplices and what fate was reserved for them? It was this disqualification of the survivors from the Stalinist team, this challenging of the system as a whole, that the secret report had, by its silences, to avoid. So that these questions would not be asked, a communist party which by definition had to explain historical phenomena in considering the totality of the conditions which accompanied them had preferred to reduce its analysis to the wrong-doing caused by flaws in the personality of one individual. In this sense, Krushchev and his companions took their place in the line drawn in 1922 by Lenin who, realising the damage wrought by rapidly increasing bureaucracy, had finally opted for the explanation that all was due to the faults of Stalin and the men who made up the leadership of the Party. Togliatti has shown the weakness of such an analytic method, wholly alien to Marxist thought.

In the last resort, the secret report revealed what destalinisation had been. It was caused by two elements. At the lower level, it was the result of the total exhaustion of society. This made the wholesale maintenance of the system intolerable and, as soon as the signs of a possible change appeared, caused a stifled hope to raise its head and animate that society. At the top, those who held power and were fighting for the succession, and those cadres of the intermediary ranks who wanted finally to guarantee their status and their privileges, also wanted to save the system of which they were the beneficiaries and to change it so that it became less destructive than it was and less intolerable to the rest of the country. The desire to change the Party existed in a precise framework. It was a question of changing the system in order to save it and not to destroy it. This meant that the Party must never lose the initiative and control over the changes; that it must not allow society to influence the process. Delivered from Stalin – who was blamed exclusively for the past errors – the Party in a return to Leninist legitimacy was to try to find a new legitimacy. This was in fact one of the problems confronting Stalin's successors. The moment they did not enthrone a new Stalin who would inherit his legitimacy, the moment they condemned the forms taken by Stalin's power from 1934 to 1953, they found themselves having to face Stalin's usurpation of legitimacy and a vacancy in legitimacy after him.

By returning to Lenin's legitimacy, by reattaching itself to the sources of the regime and to its options – because until 1934 everything had been normal – Stalin's successors sealed off the years of terror and credited the Party and its representatives with everything which had been accomplished by the USSR since the Revolution; they also reaffirmed by so doing the infallibility of the Party, which was the basis of the legitimacy of the power which it exercised. In this revision, the Communist Party of the USSR remained true to the way it saw its relations with society. As always, the Party substituted itself for society. Just as it had seen social change as a revolution from above, so it practised *destalinisation from above*.

In spite of these limitations, the considerable effects for the USSR of the destalinisation, and in particular of the XXth Congress, should not be underestimated. Soviet society, which they wanted to isolate from the destalinisation, for the most part was aware of what was happening. And it was aware also of the omissions. Intellectual liberation had been accelerated by it. Moreover, the XXth Congress, in making public a part, no matter how restricted, of Stalinism had opened in the whole system and not only in the assessment of Stalin's role, two breaches which the Soviet powers could never plug. In denouncing the Stalinist system of terror and its mechanisms, Khrushchev's report prevented it from ever returning. Also, he had begun to lift the pall of fear which had paralysed society and forced the political system to move in the direction of an increasing legitimacy. Moreover, it destroyed, involuntarily but inevitably, the dogma of the Party's infallibility and thus shook the whole fabric of the relations of society with the authorities and their ideological justification. After the XXth Congress Soviet society knew that the power of the Party was based on its ability to resist and not on an inexorable historical law. Similarly, the Soviet system found that it had been robbed of its originality and that its authority had been made relative.

From 1929 to 1953, for nearly a quarter of a century, Stalin imposed his own unlimited power on his country. This power, and the forms it took, give Stalin a central place in the tragic cohort of tyrants whose names are remembered in history. Much has already been written about him, about his passion to dominate, his utter unscrupulousness, his cruelty and at the moment nothing can be added to these analyses, especially to the astonishing portrait drawn by Boris Suvarin in 1935. Also, it is not Stalin the man whom we are trying to describe here, but the system which he established. Impelled by a passion for power, Stalin was able to satisfy it and to maintain himself for his whole lifetime at the head of the authoritarian system which he had created. But as one of his biographers, Robert C. Tucker, has very rightly remarked, Stalin cannot be reduced solely to his passion for domination because his domination emerged into the creation of a genuine system, *Stalinism*. This was, in fact, a decisive historical debate. The years of Stalinist power were, like those of Nero, for example, years of unadulterated tyranny, the course of which had been determined by the temperament of the tyrant. Were they therefore linked to the person of the tyrant and destined to disappear with him? Or was Stalinism a coherent universal system, with its own logic, capable of surviving the tyrant? If one admits the existence of a veritable Stalinism, another question immediately arises. Was it a temporary phenomenon, local and specific to Russia, the causes of which can be clearly seen in the terrain in which this phenomenon developed? Or was it a complex system, rooted undoubtedly in Russia through many of its characteristics, but also made up of external contributions and therefore with a universal significance? In other words, was it a perversion explained by the history and specificity of Russia? Or a Russian variant of contemporary totalitarian systems?

Here we suggest that a global system called Stalinism exists; that this system has two components; a theory and a practice of power on the one hand, a plan of radical change of societies on the other; that, in the end, it is the combined product of Western Marxism and the Russian tradition.

One of the components of Stalinism, the one by which it is above all defined is the system of power. Officially and to give itself legitimacy, Stalinism was defined by appealing to Marx's concept of the *dictatorship of the proletariat*. This dictatorship, in Stalin's political system, was replaced by the total power of one man, supported by several apparatuses: Party, State, police, the last of which dominated the others and was the real instrument of authority. Thanks to this instrument, Stalin governed his country by forcing it to undergo an almost constant terror which broke down the will of everyone and atomised society, leaving him faced only with the fears and weaknesses of individuals. Is this a unique system without any equivalent in history? No. Aristotle had already described tyrannies and the means of perpetuating them. For him, the tyrant could not allow his subjects to be free to think for themselves, to have freedom of action in any area, nor the possibility of living in a climate of reciprocal confidence. Is this not the definition of the modern totalitarian system which can be described by the combination of the following characteristics: an official ideology which excludes any other ideology and removes from the citizens the freedom to think outside it; a single party which holds the monopoly of decision and action; the grip of the State on all the armed forces; a State monopoly of the media and of all the means of diffusing ideas; finally, a terroristic police control. All the components which political theory from Aristotle to Brzezinsky has carefully described are to be found in the Stalinist system. But at the same time it is complicated by other traits which make it sometimes difficult to interpret. In a book refulgent with lucidity and intelligence, Hannah Arendt has underlined these characteristics. The absolute power of the past leader, in the modern epoch, characterised by the proliferation of the bureaucracy, by a chain of separate apparatuses (in the USSR the State, the Party and the police). However, between these apparatuses there is no equality. Those who hold the visible power are those who are invested with a limited power. The Soviets, the cornerstones of the Soviet constitutions of 1924 and 1936 possessed no real power; the Party which should have guided society had been decimated and infiltrated by the police. As for the latter, whose role in law was subordinate to the other apparatuses, it enjoyed a power that was all the more unlimited in that it was secret. This unlimited power of the police in a totalitarian system has, however, one limit which is the person and the security of the tyrant which governs by relying on it, but also always by being careful to contain its power. This concern explains why, although the police under its various names had been omnipotent under Stalin, its leaders had, one after the other, been liquidated just as they themselves had liquidated the political cadres. Even Beriya who had lasted longer than any of his predecessors in that office, had become, at the beginning of the 1950s, a question-mark for Stalin and was destined for a rapid fall from which only Stalin's death saved him for the time being.

Another characteristic of Stalinism, which one finds in other totalitarian systems, is the persistence and even the search for an apparent legality, although despotism and illegality are at the heart of the functioning of the system. The greater the terror, the greater the insecurity of the citizens, the more the power tries to build a façade of legality which reassures the external world and disarms the citizens, conscious, however, of the distance between the terrorist reality and the law which in appearance is triumphant. Hitler operated in this way while he officially maintained the Weimar Constitution, and Stalin also by multiplying the codes and drawing up a constitution at the very moment when a purge was decimating his country. It is significant that the same man – Vyshinsky – was at that time the great lawyer and constitutionalist of the USSR and its Supreme Procurator, embodying at the same time law and despotism.

Considered in this way, Stalinism is in no way specific to Russia. The main characteristics of this political system were also repeated at the time in Hitler's Germany and the personalities, however different, of the two despots, rooted in historical and cultural structures which it is not easy to compare, led the two dissimilar societies to parallel destinies.

But the difference begins with the plan for the future which gives Stalinism its originality. Because Stalinism was also an unprecedentedly vast plan for the modernisation of Russia, to which Stalin applied methods which owed much to his temperament. In this project Stalin took his place in the double line of Russian history and of Marxism. Russian history since the eighteenth century had been dominated by the argument about the roads to the modernisation of the country and by the attempts made to bring this about. But Marxism, in which the Bolsheviks claim to believe, and Stalin was a Bolshevik, demanded that societies should be totally transformed, starting with the revolution of the economic structures; that these changes are at once realisable and inevitable; that they are the fruit of a historical necessity and of the organised effort of a privileged agent of history. Stalin had thus inherited from both sides this idea of the total revolution of societies; and it was therefore logical that in his mind the task of the power which had emerged from the Revolution should be accompanied by this total revolution.

In 1917, Lenin, before him, had thought in terms of modernisation, of the revolution of the whole social culture and hoped that in the void left by the disappearance of the structures of the *ancien régime* the change would come of itself. He had successively applied two approaches to this change. The violent approach of War Communism. Then when he had realised the capacity of the masses to resist his project, he changed to the more gradual and moderate approach of the NEP. In 1929, Stalin in turn tackled the problem, and the answer which he brought to it was to leave a lasting and tragic mark upon the history of his country. Here it must be stressed that the revolution in which Stalin engaged in 1929 was not the fruit of temporary circumstances. The economic difficulties of the end of the decade of the 1920s are not enough to explain his turn to the left. Although some of the difficulties did exist, although they influenced some choices, it is clear that in 1929 Stalin, having rid himself of his opponents, considered that at last he had the opportunity to accomplish a vast social revolution and that this total revolution was for him the final result of the whole of Russian history, just as it had been the final outcome of 1917. Stalin was to apply to this overall project three master ideas. For him such a profound change could not come about from below because it ran counter to the social conscience. This implied that the change had to take on new forms and a cataclysmic rhythm. Besides, in his mind it was a *political* project, which is why he was indifferent to the economic conditions which would enable it to be achieved and to social consensus. Stalin's vision being essentially political, it was power which for him was the instrument of such a design. Finally, Stalin considered that all the goals which constituted the process of modernisation – the strengthening of power, the transformation of the economy, the transformation of the social, the intellectual and moral change and social integration – goals which generally show themselves and are reached separately in the course of different periods and according to different means, must constitute a whole, a single historic moment.

Total revolution, immediate, simultaneous, this plan was to clash with those who were its object and could not be brought about with them but could only be imposed upon them. From the first Stalin discerned the idea of a transformation which would associate those interested in the change and effected a *revolution from above*; he justified the constraint exercised upon society by affirming that it was in its own interest. In this, he was the heir of Leninist voluntarism, but also of the preceding attempts at modernisation of Russia, such as those of Peter the Great who made the unified State the cornerstone of all change. This idea lent a particular originality to the Stalinist conception of power and of the State. At the beginning of the 1930s, when he had overcome his adversaries, Stalin clearly modified the justification for the power which he possessed. Until then, he had presented himself above all as the arbiter between the factions in the Party which was torn apart by the quarrel over Lenin's successor. At the moment of the general change, he claimed for himself a more traditional power, a *legitimacy inherited from Lenin*. This is the important turning-point in relation to the revolutionary ideas. In 1917 the power of the Bolsheviks was born of a break with the whole past. It held its legitimacy in principle from its accord with historical necessity, from a consensus of the masses. Stalin, on the contrary, came back quickly to the idea of the *continuity* of power – he was Lenin's successor – and through this he freed himself from the demands, even if purely theoretical, of a popular consensus.

This traditional idea of power was strengthened by the fact that the State progressively became the natural structure of human activities; a State which found again its traditional attributes: its own sovereignty, a territory, a history, and thus a national dimension. It was not only the attributes of a State which reappeared at that time but also its main function, inherited from the Russian State and further back still from the Mongol tradition. The State's task was not to reflect society but to mould it. It was the State that assumed all the tasks which had hitherto been carried out by social groups; it directed, educated, defined morals and meted out justice; it was also the sole entrepreneur. This conception of power which underlies the *revolution from above* was very close in fundamentals to the conception of the Imperial State. The three key expressions which describe the Imperial State are *samoderzhaviye, pravoslaviye, narod-*

nost (autocracy, orthodoxy, nationality). If one substitutes *commun-ism* for *orthodoxy* as the unique system of values of the State and society one arrives at the definition of the Stalinist State as it had existed since 1936. Like the former State, the Stalinist State – which was identified with Stalin – did not allow any mediating structures between society and itself. But unlike the former State in which the authority of the sovereigns was progressively under-mined and weakened by the élites, Stalin's authority was to be total until the end. This power, which he had brought to a degree of extreme cohesion and power, Stalin used to mould a new society, a project which is to be found in all Utopias and which, when it passes from dream to reality, transforms in general the destiny of the peoples into a collective tragedy.

Stalin's determination to remodel society had a double *raison d'être*. The immediate one, as befitted a Marxist project, was to create a workers' society adapted to a rapid industrialisation. The more profound one was to destroy the differences which had always existed in Russia between the two worlds: the State and the largely peasant society; between the two cultures, the official culture of the State and the traditional culture of the peasants. The target of this project, whether one looks at it in its immediate dimension or in its historic dimension was always the peasantry which was in fact the target of all the projects for modernisation. Everywhere, mod-ernisation meant the movement of men towards the cities where they would take part in the industrial revolution. In Russia, this movement was all the more necessary, thought Stalin, in that it alone could subjugate society to the State and impose on society the Marxist culture which the State demanded. Thus Stalin was faced with the problem which all the leaders of that country had encoun-tered before him. The political ideas of Russia, its official political culture, had always in essentials been borrowed from abroad. Those of the Varegues coming from the north, then the Mongol occupiers who for nearly three centuries had maintained their hold on Russian history, finally those which Catherine the Great had tried to borrow from Western Europe. The Revolution of 1917 had not been omitted from this tradition, since the Bolsheviks had im-ported into their country the ideas which Marx had developed by observing the development of European societies and the Revolu-tion had been for them a decisive stage in the integration of Russia into the Western cultural world.

But, confronted by the political culture of the government, the Russian peasant society had always taken refuge in another culture which rejected foreign contributions and which was never enriched by them. The 'horizontal' culture of Russia is a profound reality based on three essential political characteristics which were rightly discerned by Bakunin: a latent anarchism, a spirit of sporadic revolt, a total alienation from the State. For the Marxists, this duality of cultures was merely the reflection of the classic opposition between the culture of the ruling classes and that of the dominated classes which could be solved by the accession to power of the oppressed. The 1920s had proved that the problem was not so simple and that the culture of the peasant masses remained alien to the Marxist political culture. The transforming project of Stalin wanted first of all to respond to this cultural dichotomy which it intended to suppress by suppressing the peasantry. To do so, he made use of the known means – he forced the peasants to emigrate to the cities, the building sites and the camps – but he also used less apparent ones, as R. C. Tucker emphasised. Stalin imprisoned the peasants in the kolkhozes where their work, their contribution to economic development passed for a genuine remission in the *corvées* (the hours of work due to the kolkhoz). He tied them to this place of obligatory work by refusing them the internal passports which were restored by a decree in 1932 in order to control the movements of Soviet citizens. In a society placed under constant surveillance, the status of the peasants appeared to be a differentiated status in which the kolkhoz, its restraints and its links, renewed the essence of the status of serfdom which had been suppressed in the Russian Empire in 1861.

At the other extreme of society, the bureaucracy with its privileges reconstituted the class of the civil servants upon which the functioning of the Empire had depended. When the police, a particularly privileged stratum of the bureaucracy, acquired absolute control of the prison labour force (mainly peasants whom they could throw into prison whenever their needs demanded it), the old link between the services rendered by the bureaucratic class and the serfs whom their sovereigns gave to them in payment for their services was completely re-established. This reconstituted serfdom was legitimated by Stalin by appealing to the ideology. In a country in which the working class is in power – the dictatorship of the proletariat has shown this to be true – the peasantry is the enemy; its culture which preaches individual values – attachment to private property and the tight community of the village – is incompatible with the ideology of the working class. Stalin's social revolution thus fell in with the efforts of his predecessors to integrate society around a single political culture which was the channel of power. But it would be too easy, and inaccurate, to reduce the Stalinist enterprise to a pure and simple prolongation of the Russian tradition. Two characteristics here come from Marxism. In the first place, the industrial ideology, the will towards economic progress and the link between industrial progress and social change. Secondly, Stalin borrowed from Marxism the certainty of following a historically inexorable path and of being justified by history. Where his predecessors had argued about what path to take and by their hesitations had humanised their enterprise, Marxism, a Utopia petrified by Marx's successors into a series of scientific certainties opened the way to continued constraint since science guaranteed that there was no alternative.

In forging Stalinism in this way – political system, modernising project – what part of himself, what part of his own temperament had Stalin injected into it? Apart from the most negative traits of his nature which his contemporaries had soon discerned – the extreme brutality, the suspicion, the cynicism – Stalin had probably stamped his creation with three aspects of his personality which time was to reveal. In the first place his passionate concern for the State. Stalin wanted to be a statesman, founder of a State and in order to be so he restored the power of the State. Conceived first as a way to modernise, the State rapidly became an end in itself and the modernisation served in the last resort to increase its power and to justify its continuance. The choice of *socialism in one country* was not accidental in Stalin; it was linked to this will to govern which ran through the quarter of a century dominated by him. This determination to favour the State also explains the Stalinist interpretation of the world revolution. The revolution for him was not an autonomous dynamic, and Stalin only acknowledged it through the Soviet State subjected to the interests of that State and prolonging them. The revolutions which he organised after the Second World War were all conceived on the Russian model – revolution from above – and were placed under the control of the Soviet State and their effect was to extend the sphere of international power of that State. How can this State-controlled vision of the Revolution be explained, if not by the fact that Stalin wanted to take his place in the revolutionary pantheon and in history? In the Revolution he had taken second place to Lenin and behind the other Bolsheviks. He had gradually eliminated from the history of his country and his Party all those who had taken part in the Revolution and rewritten in his own way an epic in which there continued to exist only, and equally, the founder of the revolutionary ideology, Marx, the man of the Revolution, Lenin, and the true founder of the communist State, Stalin. In claiming for himself the State, in identifying it with himself, in identifying the Soviet State and the Revolution, Stalin assigned to himself an unjustified place in history and gave legitimacy to an unlegitimated power.

A second Stalinist characteristic was the constant determination to preserve a revolutionary legitimacy in his practice. Holder of a total power which he could direct as he pleased, contemptuous of all ideology, Stalin nevertheless wanted to be in tune with Marxism and the ideals of the Revolution. He always justified his practice by creating theories which he claimed were Marxist. He thus went from the identification of socialism in one country to a world revolution which justified the concentration of Stalinist interest on the USSR to the detriment of the external foreign revolutions; the same is true for the theories of *capitalist encirclement* and of the *strengthened spirit of aggression of the internal and external enemies of the USSR towards the advance of a socialism*, theories which justified the strengthening of the State by the constant growth of the repressive system. The ideological justification of Stalinist practice took into account the existence of a double language in the USSR, that of the authorities which described a non-existent reality distorted in order to give itself legitimacy, and the other, for a long time muted, of a society which resented the gap between what the official language expressed and reality. In isolating the USSR from the rest of the world, for that was one of the characteristics of Stalinism which is not found in any non-communist totalitarian system, Stalin was able to make those he governed believe that the description of reality corresponded to reality. But it only needed the war and the millions of prisoners and deportees for this illusion to be shattered. Although, even in 1953, the ideology could still pretend that it reflected and justified the real, this was because the terror which had begun again condemned that society to keep its doubts to itself. But these doubts existed and they marked the limits of loyalty to the ideology. It remains, however, that this unremitting determination to justify a heterodox practice and reattach it to Marxism, was not the fruit of an inexplicable obsession alone. In Stalin's cynicism credit must also be given for a loyalty, ill-conceived but indubitable, to the ideology of the Russian Revolution.

Finally, Stalin was the product of a civilisation of confines, hesitating between Westernisation and attachment to its Russian roots;

and this ambivalence, present in his personality, appeared in his choices. Stalin retained from Marxism a desire to westernise his country, to make it follow the historical path of western societies. His almost visceral hatred of the peasants and the violence with which he treated them was a sign of this desire. Like the Westernisers before him, he wanted to tear his country away from what Lenin called 'Asiatic barbarism' and his model of modernisation was directed at this end. But at the same time, Stalin had inherited from the Slavophils a deep suspicion of the West, and this suspicion impelled him to try to find in his country, and only there, the seeds of socialism and the means to build it without the Western world. Swinging continually between these two poles, he ended with an incomplete formula which tried to reach modernisation without Westernisation and which in the last resort was a mixture of industrial progress and barbarism. His work led the historian Gerchenkron to make the disillusioned statement that in Russia attempts at modernisation have usually led to a renewal of barbarism.

The death of any tyrant marks for the system he has dominated the opening of a period of uncertainty. Can this system survive without the man who embodied it? Can it be transformed without bringing upheavals and dramas which will add still more to the sufferings already endured by the society which had until yesterday been subjected to tyranny? Stalin's death posed this classic question in particularly sharp terms because his legacy was particularly complex. It was a contradictory legacy, made up of apparent successes and profound weaknesses. In the name of Stalin, his successors could assume the power of their country. The USSR, alone, rejected by all the nations, and backward at the end of the 1920s, had become the second-largest military and economic power in the world which had consolidated its frontiers, recovered the territory lost by the Empire and extended its domination into the very heart of Europe. Stalin, in the name of Marx, had realised the old Pan-Slav dream of the Russian sovereigns. It was Stalin, too, whom his adversaries had accused of neglecting the world revolution who had helped it to make its greatest advances. Here lies one of the paradoxes of revolutionary history. So long as internationalists like Lenin and Trotsky had dominated the international communist movement and possessed a revolutionary instrument in the Comintern, the Revolution had not crossed the Russian frontiers. Conversely, when Stalin, who was centred on Russia, was in power, had destroyed the revolutionary instrument created by Lenin, he extended the field of the Revolution to Eastern Europe and to Asia. Having taken power in a *besieged fortress*, Stalin extended it so that the world outside the USSR was in a state of siege.

The extreme weaknesses of society corresponded to the extreme power of the Soviet State. The successes achieved in the economic field and to win the war had been achieved at the cost of tens of millions of human lives. This was a tragic tribute due not to a clear historical necessity but to Stalin's contempt for his fellow human beings and to his mistakes. Although in 1953 the system of Stalinist terror still imposed silence on those whom he governed, already the clear-sighted were aware that the advances had not been made because of the price paid but in spite of it, and above all in spite of Stalin's mistakes.

The spectacle of a dislocated society, exhausted by fear and privations was added to the lives lost. This new society was characterised by heterogeneity and differences, and not by unity. The bewildered and amorphous masses were opposed by the cohort of the bureaucrats, or by all those who, at two different levels served the system directly and enjoyed its privileges: political cadres, beneficiaries of the *nomenklatura* (all the posts in the gift of the Party and the holders of which represented the Soviet élite in all fields) holders of some fragment of power. Soviet society in the 1950s also contained a genuine ruling class which was characterised both by the privileges which it enjoyed and by insecurity. Stalin's absolute authority meant that at any time the positions acquired could be lost. Undoubtedly the war and the period of reconstruction had stabilised society for a time and guaranteed a certain continuity in the advantages which had been won. But everyone knew from past experience that continuity was not the dominant feature of Stalinism. The determination of the Soviet bureaucracy to perpetuate itself, to put an end to the continual renewals which destroyed it, was one of the essential moral traits which characterised it. The revolutionary enthusiasm had long since disappeared in the USSR. The war and the temporary opening to the outside world had brought a change in the convictions imposed upon society. The Soviets had become aware that socialism was not paradise and that hell did not begin at their frontiers. Without doubt, the renewed terror prevented them from expressing these insights, and the cloak of silence which fell once more on Soviet society in 1946 could pass for a total commitment to the system. But in the depths of the social conscience, the weakening process had already begun. Deprived for nearly a quarter of a century of its collective memory, the

Soviet people sensed after the war that the truth presented by those who governed them was far from corresponding to reality. This gradual disenchantment of the masses, disenchantment in the literal sense, that is to say when the conscience begins to shake off the ascendency which has been exercised over it, joined to the despairing determination of the privileged to escape at last from their precarious status inherent in Stalinism, gave to the last days of the system a special tonality. The Soviet historian Alexander Nekrich emphasised this in his memoirs as have many others: the atmosphere as Stalin's reign came to an end was intolerable, because everyone felt that the system was about to fall back into the worst excesses, or that it must be destroyed. Stalin's death closed the first course and led his potential successors to look for means by which the second could be avoided. Having learnt from past experience, all those who could influence events tried to prevent both a new tyranny because they all feared each other, and chaos because they feared the uncontrolled reactions of society. To bar the way to a new Stalin, to safeguard the political system by adapting it, to neutralise the masses and keep them outside the ineluctable process of change were the goals of the Soviet political class between 1953 and 1956. These goals concerned above all the political system and the relations with those it administered. They left in doubt the fundamental problem of Stalinism, which was one of lies and of the terror by which the lie was destined to be turned into the truth. Just as the system based on the lie and on terror had been brought to a very high degree of perfection by the determination of one single human being, Stalin, the breaches opened in this system were to be the fruit of the intuitions and the temperament of another single human being, Khrushchev. One of the paradoxes of the USSR is that of the decisive role played in its history by individuals, although the ideology of that country has proclaimed for more than sixty years that history is above all the product of the collective wills and the movements of the masses.

APPENDIX
MEMBERSHIP OF THE RULING ORGANS OF THE PARTY
1917–1953

1917
VIIth CONFERENCE (April)

CENTRAL COMMITTEE

*Members**
G. F. Fedorov
L. B. Kamenev, *executed† in 1936*
V. I. Lenin, *died on 24 January 1924*
V. P. Milyutin, *disappeared during the purges*
V. P. Nogin
I. T. Smilga, *disappeared during the purges*
J. V. Stalin, *died on 5 March 1953*
I. M. Sverdlov, *died in March 1919*
G. I. Zinovyev, *executed in 1936*

Candidates
A. S. Bubnov, *disappeared in 1938, rehabilitated, alive in 1956*
N. P. Glebov-Avilov
A. Pravdin
I. A. Feodorovich

VIth CONGRESS (August)

CENTRAL COMMITTEE

Members
I. A. Berzin
Bubnov
N. I. Bukharin, *executed in 1938*
F. E. Dzherzhinsky, *died on 20 July 1926*
Kamenev
A. M. Kollontai , *died in 1952*
N. N. Krestinsky, *executed in 1938*
Lenin
Milyutin
M. K. Muranov
Nogin
M. S. Uritsky, *died in 1918*
A. I. Rykov, *executed in 1938*
F. A. Sergeyev
S. G. Shaumyan, *died in 1918*
Smilga
G. I. Sokolnikov, *disappeared during the purges*
Stalin
Sverdlov
L. D. Trotsky, *assassinated in 1940*
Zinovyev

* The initials are only given once when the names appear for the first time in the ruling cadres.
† The word 'executed' is used for executions after a trial. The word 'liquidated' is used for executions without any known or public trial.

Candidates
P. A. Dzhaparidze
A. S. Kiselev
G. I. Lomov, *liquidated in 1937*
I. A. Preobrazhensky, *liquidated in 1938*
N. A. Skrypnik, *committed suicide in 1933*
E. D. Stasova
V. N. Yakovleva
A. A. Yoffe, *committed suicide in 1927*

1918
VIIth CONGRESS (March)

CENTRAL COMMITTEE

Members
Bukharin
Dzherzhinsky
Krestinsky
M. M. Lashevich, *died in 1928*
Lenin
V. V. Schmidt, *disappeared during the purges*
Sergeyev
Smilga
Sokolnikov
Stalin
Stasova
Sverdlov
Trotsky
M. F. Vladimirsky
Zinovyev

Candidates
Berzin
Lomov
G. I. Petrovsky, *purged, died in 1958*
A. G. Shlyapnikov, *disappeared in a labour camp*
P. Stuchka
Uritsky
Yoffe

1919
VIIIth CONGRESS (March)

I. CENTRAL COMMITTEE

Members
A. G. Beloborodov, *disappeared during the purges*
Bukharin
Dzherzhinsky
M. I. Kalinin, *died on 3 June 1946*
Kamenev
Krestinsky
Muranov
K. B. Radek, *disappeared in a labour camp*
Kh. G. Rakovsky, *disappeared in a labour camp*
L. P. Serebryakov, *executed in 1937*
Smilga
Stasova
Stuchka
M. P. Tomsky, *committed suicide on 22 August 1936*
Trotsky
G. I. Yevdokimov, *executed in 1936*
Zinovyev

Candidates
Bubnov
K. Danishevsky
A. Mitskevich-Kapsukas
Sergeyev
Schmidt
I. N. Smirnov, *executed in 1936*
M. I. Vladimirsky
I. M. Yaroslavsky, *died in 1943*

II. POLITBURO

Members
Kamenev
Krestinsky
Lenin
Stalin
Trotsky

Candidates
Bukharin
Kalinin
Zinovyev

III. SECRETARIAT

Krestinsky

1920
IXth CONGRESS (April)

I. CENTRAL COMMITTEE

Members
A. A. Andreyev
Bukharin
Dzherzhinsky
Kalinin
Kamenev
Krestinsky
Lenin
Preobrazhensky
Radek
Rakovsky
I. E. Rudzutak, *executed in 1937*
Rykov
Serebryakov
Sergeyev
Smirnov
Stalin
Tomsky
Yevdokimov
Zinovyev

Candidate members
Beloborodov
S. I. Gusev
Milyutin
V. M. Molotov
Muranov
Nogin
Petrovsky
I. A. Pyatnitsky, *disappeared in
   1937*
Smilga

Stuchka
Yaroslavsky
P. A. Zalitsky, *disappeared during
   the purges*

II. POLITBURO

Members
Kamenev
Krestinsky
Lenin
Stalin
Trotsky

Candidate members
Bukharin
Kalinin
Zinovyev

III. ORGBURO

Krestinsky
Preobrazhensky
Rykov
Serebryakov
Stalin

IV. SECRETARIAT

Krestinsky
Preobrazhensky
Serebryakov

1921
Xth CONGRESS (March)

I. CENTRAL COMMITTEE

Members
Bukharin
Dzherzhinsky
M. V. Frunze, *died 13 October 1925*
N. P. Komarov
A. Kutuzov
Lenin
V. M. Mikhaylov
Molotov
Ordzhonikidze

Stalin

Petrovsky
Radek
Rakovsky
Rudzutak
Rykov
Sergeyev
Shlyapnikov
Stalin
Tomsky
Trotsky
I. Tuntul
K. I. Voroshilov
Yaroslavsky
Zinovyev

Candidates
V. I. Chubar, *disappeared in 1938*
Gusev
S. M. Kirov, *murdered 1 December 1934*
Kiselev
V. V. Kuibyshev, *died 25 January 1935*
Milyutin
V. V. Osinsky, *disappeared in 1938*
G. L. Ryatakov, *executed in 1937*
G. I. Safarov, *disappeared in 1938*
Schmidt
Smirnov
D. I. Sulimov
N. A. Uglanov, *disappeared in 1938*
Zalitsky
I. A. Zelensky, *executed in 1938*

II. POLITBURO

Members
Kamenev
Lenin
Stalin
Trotsky

Zinovyev

Candidates
Bukharin
Kalinin
Molotov

III. ORGBURO

Members
Andreyev
Dzherzhinsky
Kuibyshev
Molotov
Rykov
Stalin
Tomsky

Candidates
Kalinin
Rudzutak
Zelensky

IV. SECRETARIAT

Stalin, *General Secretary in 1923*
Kuibyshev
Molotov

1923*
XIIth CONGRESS (April)

I. CENTRAL COMMITTEE

57 Members

II. POLITBURO

Members
Kamenev
Lenin
Rykov
Stalin

* After the XIIth Congress, the Central Committee continued to increase its membership, 57 members in 1923 (40 members, 17 candidates); 85 in 1924 (50 members, 35 candidates); 106 in 1925 (63 members, 43 candidates); 121 in 1927 (71 members, 50 candidates); 133 in 1930 (70 members, 63 candidates). Power shifted towards smaller bodies, which is why the figures for the membership of the Central Committee for these years have been omitted.

Tomsky
Trotsky
Zinovyev

Candidates
Bukharin
Kalinin
Molotov
Rudzutak

III. ORGBURO

Members
Andreyev
Dzherzhinsky
Molotov
Rudzutak
Rykov
Stalin
Tomsky

Candidates
Kalinin
Zelensky

IV. SECRETARIAT

Stalin, *General Secretary*
Molotov
Rudzutak
Zelensky

1924
XIIIth CONGRESS (May)

I. CENTRAL COMMITTEE

85 Members

II. POLITBURO

Members
Bukharin
Kamenev
Rykov
Stalin
Tomsky
Trotsky
Zinovyev

Candidates
Dzherzhinsky
Frunze
Kalinin
Molotov
Rudzutak
Sokolnikov

III. ORGBURO

Members
Andreyev
Bubnov
A. I. Dogadov
L. M. Kaganovich
Kalinin
Molotov
K. I. Nikolayeva
Uglanov
Smirnov
Stalin
Voroshilov
Zelensky

Candidates
N. K. Antipov
N. P. Chaplin
Frunze
Lepse

IV. SECRETARIAT

Stalin, *General Secretary*
Andreyev
Kaganovich
Zelensky

1925
XIVth CONGRESS (December)

I. CENTRAL COMMITTEE

106 Members

II. POLITBURO

Members
Bukharin

Kalinin
Molotov
Rykov
Stalin
Tomsky
Trotsky
Voroshilov
Zinovyev

Candidates
Dzherzhinsky
Kamenev
Petrovsky
Rudzutak
Uglanov

III. ORGBURO

Members
Andreyev
A. B. Artyukhina
Bubnov
Dogadov
S. V. Kosior, *liquidated in 1938*
E. I. Kviring
Molotov
Smirnov
Stalin
Uglanov
Yevdokimov

Candidates
Chaplin
Lepse
Mikhaylov
Schmidt
K. V. Ukhanov

IV. SECRETARIAT

Members
Stalin, *General Secretary*
Kosior
Molotov
Uglanov
Yevdokimov

Candidates
Artyukhina
Bubnov

1927
XVth CONGRESS (December)

I. CENTRAL COMMITTEE

Trotsky, Zinovyev and Kamenev
were not re-elected

II. POLITBURO

Members
Bukharin
Kalinin
Kuibyshev
Molotov
Rudzutak
Rykov
Stalin
Tomsky
Voroshilov

Candidates
Andreyev
Chubar
Kaganovich
Kirov
Kosior
A. I. Mikoyan
Petrovsky
Uglanov

III. ORGBURO

Members
Andreyev
Artyukhina
Bubnov
Dogadov
Kosior
N. A. Kubyak
Molotov
I. M. Mozkvin

M. L. Rukhimovich
Smirnov
Stalin
Sulimov
Uglanov

Candidates
V. A. Kotov
Lepse
Lobov
Mikhaylov
Schmidt
Ukhanov

IV. SECRETARIAT

Members
Stalin, *General Secretary*
Kosior
Kubyak
Molotov
Uglanov

Candidates
Artyukhina
Bubnov
Mozkvin

1930
XVth CONGRESS (July)

I. CENTRAL COMMITTEE

Bukharin, Tomsky and Rykov
*remain members of the Central
Committee (CC)*

II. POLITBURO

Members
Kaganovich
Kalinin
Kirov
Kosior
Kuibyshev
Molotov
Rudzutak

Rykov
Stalin
Voroshilov

Candidates
Andreyev
Chubar
Mikoyan
Petrovsky
S. I. Syrtsov, *disappeared in 1937*

III. ORGBURO

Members
K.I. Bauman
Bubnov
I. B. Gamarnik, *committed suicide in
1937*
Kaganovich
Lobov
Molotov
Mozkvin
P. P. Postyshev
Shvernik
Stalin
An unidentified member

Candidates
Dogadov
A. V. Kosarev
Smirnov
A. M. Tsikhon

IV. SECRETARIAT

Members
Stalin, *General Secretary*
Bauman
Kaganovich
Molotov
Postyshev

Candidates
Mozkvin
Shvernik

Stalin

1934
XVIIth CONGRESS (February)

I. CENTRAL COMMITTEE

Of 139 Members
71 members – 68 candidates

II. POLITBURO

Members
Andreyev
Kaganovich
Kalinin
Kirov
Kosior
Kuibyshev
Molotov
Ordzhonikidze
Stalin
Voroshilov

Candidates
Chubar
Mikoyan
Petrovsky
Postyshev
Rudzutak

III. ORGBURO

Members
Gamarnik
Kaganovich
Kirov
Kosarev
Kuibyshev
Shvernik
Stalin
A. I. Stetsky
N. I. Yezhov, *disappeared in 1939*
A. A. Zhdanov, *died in 1948*

Candidates
Kaganovich
A. I. Krinitsky

IV. SECRETARIAT

Stalin

Kaganovich
Kirov
Zhdanov

1939
XVIIIth CONGRESS (March)

I. POLITBURO

Members
Andreyev
Kaganovich
Kalinin
N. S. Khrushchev, *member of the
    CC at the XVIIth Congress*
Mikoyan
Molotov
Voroshilov
Zhdanov

Candidates
L. P. Beriya (*elected to the CC at the
    XVIIth Congress, liquidated in
    December 1953*)
Shvernik

1952
XIXth CONGRESS (October)

I. PRESIDIUM OF THE CC

Members
V. M. Andryanov
A. B. Aristov
Beriya
N. A. Bulganin, *elected candidate
    member at the XVIIth Congress*
M. F. Chkiyatov
S. D. Ignatyev
Kaganovich
Khrushchev
A. I. Kirichenko
V. V. Kuznetsov
V. A. Malyshev
L. G. Melnikov
N. A. Mikhaylov

Mikoyan
Molotov
M. G. Pervukhin
P. K. Ponomarenko
M. Z. Saburov
Stalin
M. A. Suslov
Voroshilov

Candidates
L. I. Brezhnev
N. G. Ignatov
I. G. Kabanov
A. N. Kosygin
N. S. Patolichev
N. N. Pegov
A. M. Puzanov
I. I. Tevosyan
A. I. Vyshinsky
P. F. Yudin
A. G. Zverev

1953

PRESIDIUM, *formed on 6 March*

(The names of the members are
placed as an exception in the
order of precedence as
reproduced in the communiqué
in *Pravda*, 7.3.1953.)

Members
Malenkov
Beriya
Molotov
Khrushchev
Bulganin
Kaganovich
Mikoyan
Saburov
Pervukhin

Candidates
Shvernik
Ponomarenko

Melnikov
M. D. Bagirov

1956
XXth CONGRESS (February)

I. PRESIDIUM OF THE CC

Members
Bulganin
Kaganovich
Khrushchev
Kirichenko
Malenkov
Mikoyan
Molotov
Pervukhin
Saburov
Suslov
Voroshilov

Candidates
Brezhnev
D. T. Shepilov
Shvernik
E. A. Furtseva
G. K. Zhukov
N. A. Mukhitdinov

II. SECRETARIAT

Khrushchev (*First Secretary since
1953*) *joined the Secretariat in
1940*
A. B. Aristov, *joined the Secretariat
in 1955*
N. I. Belyayev, *joined the Secretariat
in 1955*
Brezhnev, *joined the Secretariat in
1956*
D. T. Shepilov, *joined the Secretariat
in 1955*
E. A. Furtseva, *joined the Secretariat
in 1956*
P. N. Pospelov, *joined the Secretariat
in 1953*
Suslov, *joined the Secretariat in 1947*

BIBLIOGRAPHY: ORIGINAL SOURCES

OFFICIAL TEXTS

The State

GENKINA *Obrazovaniye SSSR – Sbornik dokumentov, 1917–24* (The formation of the USSR – collection of documents), Moscow, 1949.

Istoriya sovetskoy konstitutsii v dokumentakh, 1917–56 (History of the Soviet Constitutions, documents), Moscow (Gosyurizdat), 1957, 1047 pp.

Konstitutsiya (osnovnoy zakon) SSSR, konstitutsii (osnovnyye zakony) soyuznikh sovetskikh sotsialisticheskikh respublik (The Constitution (fundamental law) of the USSR, the Constitutions (fundamental laws) of the Federated Soviet Socialist Republics), Moscow (Gosyurizdat), 1956, 492 pp.

Sbornik deystvuyushchikh dogovorov, soglasheniy i konventsiy zaklyuchennykh SSSR s innostrannymi gosudarstvami (Collection of treaties, agreements and conventions, concluded by the USSR with foreign powers), Moscow (Gospolitizdat), 1920–67, 22 vols.

Sbornik zakonov SSSR v dvukh tomakh, 1938–67 (Collection of the laws of the USSR in two volumes), Moscow (Izdatel'stvo Izvestiya Sovetov deputatov trudyashchikhsya SSSR), 1968, 752 and 896 pp.

Zasedaniye verkhovnogo soveta SSSR, Stenograficheskiy otchet (Stenographic minutes of the sessions of the Supreme Soviet of the USSR), Moscow (Izdaniye verkhovnogo Soveta SSSR), 1938, Collections published regularly since 1938.

Party

Istoriya kommunisticheskoy Partii sovetskogo soyuza (History of the CPSU), Moscow (Gospolitizdat), 1964–68, 6 vols.

Protokoly i stenograficheskiye otchety s'ezdov i konferentsiy kommunisticheskoy partii (Gospolitizdat), since 1958.

Istoriya Vsesoyznoy Kommunisticheskoy Partii (*bol'shevikov*), *Kratkiy kurs* (Short history of the CPSU), Moscow (Gospolitizdat), 1938, 352 pp. Other editions appeared in 1949 (408 pp.), 1962 (784 pp.), 1970.

KPSS v rezolyutsiyakh i resheniyakh s'ezdov, konferentsiy i plenum TsK (Resolutions of the congresses and plenums of the CC since 1898), 4 vols, 7th edn, Moscow (Gospolitizdat), 1954 (3 vols) and 1960.

Three collections of basic texts published in the USA in excellent translations can be added here:

MEISEL, JAMES HANS, KOSERA, EDWARD S. (eds). *Materials for the Study of the Soviet System, State and Party Constitution. Laws, decrees, decisions and official statements of the leaders in translation.* Ann Arbor (George Wahr publishing Co.), 1950 – xii, 496 pp.

TRISKA, JAN F., SCHLUSSER, R. M. *A Calendar of Soviet Treaties, 1917–1957*, Stanford (Stanford University Press), 1959, 530 pp.

TRISKA, JAN F. *Soviet Communism: Programs and Rules* (Documents and materials published by the Party), San Francisco, 1962, 196 pp.

BUNYAN, J., FISCHER, H. H. *The Bolshevik Revolution 1917–1918* (Collected Documents), Stanford (Stanford University Press), 1961, 735 pp.

Istoriya velikoy otechestvennoy voyny sovetskogo soyuza, 1941–1945 (History of the Great Patriotic War, published by the Institute of Marxism–Leninism), Moscow (Izdatel'stvo ministerstva oborony SSSR), 1960–65, 6 vols.

Iz istorii grazhdanskoy voyny v SSSR – Sbornik dokumentov i materialov v trekh tomakh, 1918–22 (Collected documents on the Civil War published by the Institute of Marxism–Leninism of the CC), Moscow (Izdatel'stvo sovetskaya Rossiya), 1960–62, 3 vols.

Kommunisticheskaya Partiya v period velikoy otechestvennoy voyny (*iyun' 1945 goda–1945*) *dokumenty i materialy* (The CP in the Great Patriotic War), Moscow (Gospolitizdat), 1961, 704 pp.

Velikaya oktyabr'skaya sotsialisticheskaya revolyutsiya. Dokumenty i materialy (Documents and materials on the October Revolution published by the Institute of Marxism–Leninism, Central Committee of the CPSU), Moscow (Izdatel'stvo Sovetskaya Rossiya), 1960–62, 3 vols.

Trials

Trial of the 'Bloc of the Anti-Soviet Rightists and Trotskyists' before the Military College of the Supreme Court of the USSR of N. I. Bukharin, A. I. Rykov, G. G. Yagoda, N. N. Krestinsky and Ch. G. Rakovsky. Verbatim report of the debates (from 2–12 March 1938), Moscow (People's Commissariat of Justice of the USSR), 1938, 850 pp.

Trial of the Trotskyist Anti-Soviet Centre before the Military Tribunal of the Supreme Court of the USSR of Y. L. Piatakov, M. B. Radek, G. Y. Sokol'nikov, L. Serebryakov, N. I. Muralov. Verbatim report of the debates (25–30 January 1937), Moscow (People's Commissariat of Justice of the USSR), 1937, 604 pp.

BROUÉ, P. (ed.). *Les Procès de Moscou*, Paris (Julliard), 1964, 304 pp.

TUCKER, R. C., COHEN, S. F. (eds). *The Great Purge Trial*, New York, (Grosset and Dunlap), 1965, 725 pp.

Statistics

Dostizheniye sovetskoy vlasti za 40 let v tsifrakh (The achievements of the Soviet regime in the last 40 years), Moscow (Gosudarstvennoye statisticheskoye izdatel'stvo), 1957, 370 pp.

Sotsialisticheskoye stroitel'stvo SSSR (The building of socialism in the USSR), Moscow, 1936, 719 pp.

WORKS WRITTEN BY BOLSHEVIKS

BUKHARIN, N. I., PREOBRAZHENSKY, E. *Azbuka Kommunizma*, Petrograd (Gosudarstvennoye Izdatel'stvo), 1920, 323 pp.

A French edition was published by Pierre Broué with the title *l'ABC du communisme*, Paris (Maspero), 1968.

BUKHARIN, N. I. *Teoriya istoricheskogo materializma*, Moscow (Giz), 1921, 383 pp.

BUKHARIN, N. I. *The Path to Socialism in Russia, selected works – Put' k sotsializmu v Rossii – izbrannyye proizvedeniya*, New York (Omikron books), 1967, 416 pp.

KRUPSKAYA, N. K. *Vospominaniya o Lenine.* (Recollections of Lenin), Moscow (Partiizdat), 1932.

KRUPSKAYA, N. K. *Lenin i partiya* (Lenin and the Party), Moscow (Gospolitizdat), 1963, 258 pp.

KRUPSKAYA, N. K. *O Lenine, Sbornik statey i vystupleniy* (On Lenin, collected articles and statements), Moscow (Izdatel'stvo politicheskoy literatury), 1965, 400 pp.

LENIN. *Sochineniya* (Works), Moscow–Leningrad (Gosudarstvennoye Izdatel'stvo), 1927–35, 31 vols, 3rd edn. The interest of this incomplete edition of Lenin's works lies in the editorial team, which included Kamenev and Bukharin.

LENIN. *Polnoye sobraniye sochineniy*, Moscow (Gospolitizdat), 1958–65, 55 vols, 5th edn. Very complete edition of Lenin's works, with frequent interesting footnotes.

Leninskiy Sbornik (Lenin Collection), Moscow (Gospolitizdat), 1924–59, 36 vols. This collection contains various texts which are not to be found in any of the five editions of Lenin's works, letters, drafts, etc.

LUXEMBURG, R. *The Russian Revolution*, Ann Arbor (University of Michigan Press), 1961, 109 pp.

MOLOTOV, V. *V bor'be za sotsialism, stat'i i rechi* (In the fight for socialism, articles and speeches), Moscow, 1935.

PREOBRAZHENSKY, E. *Novaya ekonomika–opyt teoreticheskogo analyza sovetskogo khozyaystva* (The new economy), Moscow, 1926, 2 vols.

PREOBRAZHENSKY, E. *De la NEP au socialisme. Vues sur l'avenir de la Russie et de l'Europe*, Strasbourg (edn CNRS), 1966, 125 pp.

STALIN, J. V. *Sochineniya – 1901–1934*, Moscow (Gospolitizdat), 1936–51, 13 vols. *Works 1901–1934*, Moscow (Foreign Languages Publishing House), 1952–55, 13 vols.

STALIN, J. V. *Les problèmes économiques du socialisme en URSS*, Paris (Editions sociales), 1952.

TROTSKY, L. *Lénine*, Paris (Librairie du Travail), 1925, 231 pp.

TROTSKY, L. *Sochineniya* (Works), Moscow–Leningrad (Gosizdat), 1926–27, 21 vols.

TROTSKY, L. *Moya Zhizn'* (My Life), Berlin (Granit), 1930, 2 vols, 327 and 339 pp.

TROTSKY, L. *Permanentnaya revoliutsinya*, Berlin (Granit), 1930, 171 pp.

TROTSKY, L. *Istoriya Russkoy revoliutsii*, Berlin (Granit), 1931–33, 2 vols, 531 and 873 pp.

TROTSKY, L. *Stalinskaya shkola falsifikatsii* (The Stalinist school of falsification), Berlin (Granit), 1932.

TROTSKY, L. *Staline*, Paris (Grasset), 1948, 623 pp.

TROTSKY, L. *The Third International after Lenin*, New York, 1957, 400 pp.

TROTSKY, L. *La révolution trahie*, Paris (Quatrième Internationale), 1961, xxxii, 271 pp.

MEIJER, JAN M. (ed.). *The Trotsky Papers 1917–1922*, vol. I, 1917–19, London – The Hague — Paris (Mouton), 1964, xvi, 858 pp. These documents from the Trotsky archives contain an important correspondence between Trotsky and Lenin at the time of the Civil War.

MEMOIRS

ABRAMOVITCH, R. *The Soviet Revolution 1917–1939*, London (G. Allen and Unwin), 1962, xviii, 474 pp. Abramovitch, together with Martov, was one of the Menshevik leaders. This work is based on his recollections and on excellent first-hand sources.

ALLILUYEVA, S. *Dvatsat' pisem k drugu* (Twenty letters to a friend), New York (Harper and Row), 1969, 383 pp.

ALLILUYEVA, S. *Tol'ko odin god* (Only one year), New York (Harper and Row), 169. 383 pp.

These two books by Stalin's daughter give a unique picture by a member of the family of the head of the USSR; in spite of its heated character, it should be taken into account.

BALABANOFF, A. *Lenin visto da vicino*, Rome (Opere Nuove), 1959, 228 pp.

DENIKIN, A. I. *Ocherki Russkoy Smuty* (Essay on the Russian storm), Paris (J. Povolzki), 1921–26, 5 vols.

Etikh dney ne smolknet slava (The glory of those days will never die), Moscow (Gospolitizdat), 1958, 492 pp. Collection of memories of the Civil War published by the Museum of the Revolution.

GORBATOV, A. V. *Les années de ma vie*, Paris (Stock), 1966, 219 pp. Memoirs of a Soviet general which throw light on the composition and the mentality of the Soviet army in the 1930s up to the outbreak of the Second World War, and the purges in this closed circle.

Gosudarstvennyy muzey revolyutsii SSSR – Rasskazivayut uchastniki velikogo oktyabrya, Moscow (Gospolitizdat), 1957, 468 pp.

Memoirs on the Revolution, published by the Museum of the Revolution.

JULLIEN, CHARLES ANDRÉ. Journal de voyage en URSS, 1921, unpublished manuscript.

KERENSKY, A. *La Russie au tournant de l'histoire*, Paris (Plon), 1967, 699 pp.

KHRUSHCHEV, N. *Souvenirs*, Paris (Laffont), 1971, 591 pp. The authenticity of this work has still to be proved. Even so, it does not contain anything sensational.

KONYEV, I. S. *Sorok Pyatyy* (1945), Moscow (Voennoye Izdatel'stvo ministerstva oborony SSSR), 1966, 280 pp.

LENIN, V. I., GORKY, A. M. *Pis'ma, Vospominaniya, Dokumenty* (*Letters, Reminiscences, documents*), 1958, 432 pp. (*Izdatel'stvo Akademii Nauk*).

Lénine tel qu'il fut, recollections of his contemporaries, Moscow (edn in foreign languages, Progress editions), 1958–65, 3 vols. 759, 959, 496 pp.

MILYUKOV, P. N. *Political memoirs 1859–1917*, Ann Arbor (University of Michigan Press), 1967, xviii, 508 pp.

MORIZET. *Chez Lénine et Trotski*, Moscow, 1921, Paris (La Renaissance du Livre), 1922, xvi, 301 pp. Reminiscences of a delegate to the IIIrd Congress of the Comintern which gives a very vivid picture of the main leading Bolsheviks.

REED, J. *Ten Days that Shook the World*, New York (Modern Library), 1935 xxiv, 371 pp. With a preface by Lenin, this book is a mine of information on the October Revolution.

Reminiscences of Lenin by his relatives, Moscow (Foreign Languages Publishing House), 1959, 224 pp. This book is a collection of the reminiscences and extracts from the writings of Lenin's sisters, of his brother Dmitri and of Krupskaya.

SADOUL, J. *Notes sur la révolution bolchevique*, Paris (Maspero), 1971, 465 pp.

SALISBURY, H. *Moscow Journal, the End of Stalin*, Chicago (University of Chicago Press), 1962, 450 pp. Day-to-day life in the USSR after 1949.

SUKHANOV, N. N. *Zapiski o revolyutsii*, Berlin (Z. I. Grizhebina) 1922–23, 7 vols. Notes on the Revolution. The most lucid eyewitness account of the Revolution.

TSERETELLI, I. G. *Vospominaniya o fevral'skoy revolyutsii*, Paris (Mouton), 1963, 2 vols, 493 and 430 pp. Recollections of the February Revolution. An eloquent defence of his policy by one of the leaders in the provisional government, undoubtedly by Sukhanov.

VOLINE. *La révolution inconnue 1917–1921;* unpublished material on the Russian Revolution, Paris (P. Belfond), 1969, 696 pp. The Revolution from the point of view of an anarchist, and the struggle of Makhno in the years of the Civil War.

YEREMENKO, A. I. *Stalingrad*, Moscow (Voyennoe Izdatel'stvo ministerstva oborony SSR), 1951, 504 pp.

ZHUKOV, G. *Vospominaniya i razmyshleniya*, Moscow (Izdatel'stvo Agenstva Pechati Novosti), 1969, 752 pp. 'Recollections and reflections' of the conqueror of Berlin.

Official biographies

'*Deyateli SSSR i Oktyabrskoy revolyutsii*', Moscow, (Entsiklopedicheskiy slovar' russkogo bibliograficheskogo Instituta Granat), 7th edn, 1927–29. The leaders of the USSR and of the October Revolution. Encyclopedia containing 246 autobiographies or authorised biographies of all the leaders who played a part in the Revolution or in the years which followed.

Zhizn' zamechatel'nykh liudey seriya biografii, Moscow (Izdatel'stvo TsK. VLKSM, Molodaya Gvardiya), 1964 (The lives of illustrious men). This biographical series is devoted to the praises of the leaders of the revolution and of the period of socialism, such as Kirov, Tukhachevsky, etc. They are official biographies.

HAUPT, G., MARIE, J. J. *Les bolcheviks par eux-mêmes*, Paris (Maspero), 1969, 396 pp. Fifty biographies taken from the Granat encyclopedia, accompanied by very valuable footnotes.

A new type of work must be added to these biographies:

MEDVEDEV, R. A. *Let History judge: the Origins and Consequences of Stalinism*, New York (Knopf), 1971, 566 pp. This book is by a Soviet historian who was only able to have it published in the USSR in the form of Samizdat. The authenticity of the text is incontestable.

BIBLIOGRAPHY: SECONDARY SOURCES

POLITICAL HISTORY

ARMSTRONG, J. A. *Ukrainian Nationalism*, New York (Columbia Univ. Press), 1963, xvi, 361 p.

ARMSTRONG, J. A. (ed.). *Soviet Partisans in World War II*, Madison (Univ. of Wisconsin Press), 1964, xx, 792 pp.

ASPATURIAN, V. V. *The Union Republics in Soviet Diplomacy, a Study of Soviet Federalism in the service of Soviet Foreign Policy*, Geneva (Droz), 1960, 220 pp.

BARGHOORN, F. C. *Soviet Russian Nationalism*, New York (Oxford Univ. Press), 1956, xii, 330 pp.

BENNIGSEN, A., QUELQUEJAY, C. *Les mouvements nationaux chez les musulmans de Russie. Le 'sultangalievisme' au Tatarstan*, Paris (Mouton), 1969, 285 pp.

BETTELHEIM, CH. *La lutte des classes en URSS, 1917–1930*, Maspero, Seuil, 1974, 520 pp. Vol. II, 1977, 605 pp.

BRUHAT, J. *Lénine*, Paris (Club français du livre), 1960, 386 pp.

BRUHAT, J. *Histoire de l'URSS*, 8th edn, Paris (PUF), 1967, 128 pp. Useful introduction to a little-known history.

BRZEZINSKI, Z. K. *The Permanent Purge. Politics in Soviet Totalitarianism*, Cambridge (Harvard Univ. Press), 1956, 256 pp.

CARR, E. H. *A History of Soviet Russia*, (7 vols) London (Macmillan), 1950–64.

The Bolshevik Revolution 1917–1923, Vols 1–3, 1950–53, 430 pp.; viii, 400 pp.; x, 614 pp.

The Interregnum 1923–1924, vol. 4, 1954, viii, 392 pp.

Socialism in One Country 1924–1926, vols 5–7 1958–64, x, 557 pp.; viii, 493 pp.; xii, 1050 pp.

CARR, E. H. and DAVIES, R. W. *The Foundations of a Planned Economy – 1928–1929* (2 pts), 1969–71, xvi, 1023 pp. The most complete of the histories of the USSR, based on original sources.

CARRÈRE D'ENCAUSSE, H. *Réforme et révolution chez les musulmans de l'Empire russe, Bukhara 1867–1924*, Paris (Armand Colin), 1966, 313 pp.

CHAMBERLAIN, W. H. *The Russian Revolution 1917–1921*, New York (Macmillan), 1935, 2 vols, xiv, 511 pp. and xii, 556 pp.

COHEN, S. *Bukharin and the Bolshevik Revolution. A Political Biography, 1888–1938*, New York, 1973.

CONQUEST, R. *Soviet Nationalities Policy in Practice*, London (Bodley Head), 1967, 160 pp.

CONQUEST, R. *The Great Terror – Stalin's Purge of the Thirties*, London (Macmillan), 1968, xiv, 633 pp.

CONQUEST, R. *The Nation-Killers, Soviet Deportations of Nationalities*, London (Macmillan), 1970, 222 pp.

COQUIN, F. X. *La révolution russe*, Paris (PUF), 1962, 128 pp.

DALLIN, D. *German Rule in Russia*, London, 1957, xx, 697 pp.

DANIELS, R.V. *The Conscience of the Revolution*, Cambridge (Harvard Univ. Press), 1960, 526 pp.

DEUTSCHER, I. *Trotsky – 1879–1940*, London (Oxford Univ. Press), 1954–63, 3 vols., xii, 540 pp., xvi, 491 pp., xvi, 543 pp.

DEUTSCHER, I. *Stalin a Political Biography*, London (Oxford Univ. Press), 1967, xx, 661 pp.

DJILAS, M. *Conversations avec Staline*, Paris, 1962, 220 pp.

EASTMAN, M. *Léon Trotski: the Portrait of a Youth*, New York (Greenberg), 1935, 217 pp.

FAINSOD, M. *Smolensk under Soviet Rule*, Cambridge (Harvard Univ. Press), 1958, xii, 484 pp.

FERRO, M. *La révolution de 1917, la chute du tsarisme et les origines d'Octobre*, Paris (Aubier), 1967, 706 pp., vol. II, *Octobre, naissance d'une société*, Paris, 1976, 517 pp.

FISCHER, L. *The Life and Death of Stalin*, London (J. Cape), 1953, xii, 707 pp.

FOOTMAN, D. *Civil War in Russia*, London (Faber and Faber), 1961, 328 pp.

GOURE, L. *The Siege of Leningrad*, London (Oxford Univ. Press), 1962, xvi, 364 pp.

KEEP, J. *The Russian Revolution*, New York, 1976, 614 pp.

KENNAN, G. *Russia Leaves the War*, Princeton, 1956, 544 pp.

KOLARZ, W. *Russia and her Colonies*, London (G. Philip), 1952, xiv, 335 pp.

KOLARZ, W. *Les colonies russes d'Extrême-Orient*, Paris (Fasquelle), 1955, 238 pp.

KOSTIUK, H. *Stalinist Rule in the Ukraine, a Study of the Decade of*

Mass Terror 1929–1939, London (Stevens and Sons), 1960, xiv, 162 pp.

LALOY, J. *Le socialisme de Lénine*, Paris (Desclée de Brouwer), 1967, 316 pp.

LÉWIN, M. *Le dernier combat de Lénine*, Paris (Éditions de Minuit), 1967, 173 pp.

LINDON, C. A. *Krushchev and the Soviet Leadership, 1957–1964*, Baltimore, 1966, 273 pp.

MARIE, J. J. *Staline 1879–1953*, Paris (edn. du Seuil), 1967, 301 pp.

NAVILLE, P. *Trotski vivant*, Paris (Julliard), 1962, 200 pp.

PAGE, S. *The Formation of the Baltic States, a Study of the Effects of Great Power Politics, the Emergence of Lithuania, Latvia and Estonia*, Cambridge (Harvard Univ. Press), 1959, xiv, 196 pp.

PARK, A. G. *Bolshevism in Turkestan, 1917–1927*, New York (Columbia Univ. Press), 1957, xviii, 428 pp.

PIPES, R. *The Formation of the Soviet Union, Communism and Nationalism 1917–1923*, Cambridge (Harvard Univ. Press), 1964, xiv, 365 pp.

RADKEY, O. *The Sickle under the Hammer*, New York (Columbia Univ. Press), 1962, 525 pp.

RAUCH, G. VON, *A History of Soviet Russia* (translated from German), New York (Praeger), 1967, xiv, 530 pp.

ROSENBERG, A. *Histoire du bolchevisme*, Paris (Grasset), 1967, 359 pp.

SCHAPIRO, L. *The Origin of the Communist Autocracy. Political Opposition in the Soviet State, First Phase 1917–1922*, London (G. Bell), 1955, xviii, 397 pp.

SHUB, D. *Lenin, a biography*, Baltimore (Penguin Books), 1966, 496 pp.

SCHUMAN, F. L. *Russia Since 1917, Four Decades of Soviet Politics*, New York (Knopf), 1962, xvi, 508 pp.

SCHWARZ, S. *The Jews in the Soviet Union*, Syracuse (Syracuse Univ. Press), 1951, xviii, 380 pp.

SORLIN, P. AND SORLIN, I. *Lénine, Trotski, Staline 1921–1927*, Paris (Armand Colin), 1961, 272 pp.

SOUVARINE, B. *Staline. Aperçu historique du bolchevisme*, Paris (Plon), 1935, 534 pp.

TUCKER, R. C. *Stalin as Revolutionary, 1879–1929*, New York, 1973, 519 pp.

ULAM, A. B. *Lenin and the Bolsheviks, the Intellectual and Political History of the Triumph of Communism in Russia*, London (Secker and Warburg), 1966, x, 598 pp.

ULAM, A. B. *Ideologies and Illusions*, Cambridge Harvard Univ. Press, 1976, 335 pp.

ULAM. A. B. *Staline, l'homme et son temps*, translated from English, Paris, 1977, vol. I 536 pp., vol. II 412 pp.

VAKAR, N. P. *Bielo-Russia, The Making of a Nation, a Case Study*, Cambridge (Harvard Univ. Press), 1956, 297 pp.

WALSH, W. B. *Russia and the Soviet Union, a Modern History*, Ann Arbor (Univ. of Michigan Press), 1968, xvi, 682 pp.

WALTER, G. *Lénine*, Paris (Julliard), 1950, 543 pp.

WERTH, A. *Russia and the War 1941–1945*, London (Barrie and Rockliff), 1964, xxvi, 1100 pp.

WOLFE, B. *Three Who Made a Revolution, a Biographical History*, New York (Dial Press), 1964, x, 661 pp.

STATE AND INSTITUTIONS

BAUER, R. A., INKELES, A., KLUCKHOHN, C· *How the Soviet System Works*, Cambridge (Harvard Univ. Press), 1957, xiv, 274 pp.

BERMAN, H. J. *Justice in the USSR, an Interpretation of Soviet Law*, Cambridge (Harvard Univ. Press), 1963, x, 450 pp.

BROUÉ, P. *Le parti bolchevique*, Paris (Éditions de Minuit), 1963, 628 pp.

BRZEZINSKI, Z. K. *Ideology and Power in Soviet Politics*, New York (Praeger), 1962, vi, 180 pp.

CHAMBRE, H. *Le marxisme en Union soviétique*, Paris (edn. du Seuil), 1955, 511 pp.

CHAMBRE, H. *L'Union soviétique. Introduction á l'étude de ses institutions*, Paris (Librairie générale de droit et de jurisprudence), 1966, 240 pp.

CHAMBRE, H. *L'évolution du marxisme soviétique*, Paris (edn. du Seuil), 1974, 415 pp.

CONQUEST, R. *Power and Policy in the USSR*, London (Macmillan), New York (St. Martin's Press), 1961, x, 486 pp.

FAINSOD, M. *How Russia is Ruled*, Cambridge (Harvard Univ. Press), 1963, xiv, 684 pp.

GELARD, F. *Les organisations de masse en Union soviétique; syndicats et komsomols*, Paris (Cujas), 1965, xvi, 240 pp.

HAZARD, J. *Law and Social Change in the USSR*, London (Stevens), 1953, xxiv, 310 pp.

HAZARD, J. *The Soviet System of Government*, Chicago (Univ. of Chicago Press), 1968, x, 275 pp. (4th edn.).

HOUGH, J. F., FAINSOD, M. *How the Soviet Union is Governed*, Cambridge, Harvard Univ. Press, 1979, 679 pp.

INKELES, A., BAUER, R. A. *The Soviet Citizen*, Cambridge (Harvard Univ. Press), 1959, xx, 533 pp.

LAVROV, D. G. *Les libertés publiques en Union soviétique*, Paris (A. Pedone), 1963, 267 pp.

LEITES, N. *The Operational Code of the Politburo*, New York, 1951, 100 p.

LEITES, N. *A Study of Bolshevism*, Glencoe (Free Press), 1953, 639 pp.

LESAGE, M. *Les régimes politiques de l'URSS et de l'Europe de l'Est*, Paris (PUF), 1971, 367 pp.

MARCUSE, H. *Soviet Marxism, a Critical Analysis*, New York (Columbia Univ. Press), 1958, viii, 271 pp.

MEHNERT, K. *L'homme soviétique*, Paris (Plon), 1960, vi, 370 pp.

MEISSNER, B. *The Communist Party of the Soviet Union*, New York (Praeger), 1956, 276 pp.

MEYER, A. G. *The Soviet Political System*, New York (Random House), 1965, x, 495 pp.

MOORE, B. JR. *Terror and Progress in USSR*, Cambridge (Harvard Univ. Press), xxii, 261 pp.

NICOLAEVSKI, B. I. *Power and the Soviet Elite; 'the Letter of an Old Bolshevik' and Other Essays*, New York (Praeger), 1965, xxii, 275 pp.

RIGBY, T. H. *Communist Party Membership in the USSR 1917–1967*, Princeton (Princeton Univ. Press), 1968, xvii, 574 pp.

SCHAPIRO, L. *The Government and Politics of the Soviet Union*, London (Methuen, Eyre and Spottiswoode), 1967, 176 pp.

SCHAPIRO, L. *The Communist Party of the Soviet Union*, London (Univ. Paperbacks, Methuen & Co.), 1970, xviii, 686 pp. (2nd edn, brought up to date).

SCOTT, D. *Russian Political Institutions*, London, 1958, 265 pp.

SCHWARTZ, S. M. *Labor in the Soviet Union*, New York (Praeger), 1952, xviii, 364 pp.

ECONOMY

BAYKOV, A. *The Development of the Soviet Economic System*, London (Cambridge Univ. Press), 1946, 514 pp.

BELOV, F. *The History of Soviet Collective Farms*, New York, 1955, 237 pp.

BERGSON, A. *The Real National Income of Soviet Russia since 1928*, Cambridge (Harvard Univ. Press). 1961, xx, 472 pp.

BERGSON, A. *The Economics of Soviet Planning*, New Haven, 1964, 394 pp.

BERLINER, J. S. *Factory and Manager in the USSR*, Cambridge (Harvard Univ. Press), 1957, xviii, 386 pp.

BETTELHEIM, C. *L'économie soviétique*, Paris (Sirey), 1950, viii, 472 pp.

BETTELHEIM, C. *Problémes théoriques et pratiques de la planification*, Paris, 1966, 304 pp.

CHAMBRE, H. *L'aménagement du territoire en URSS*, Paris (Mouton), 1959, 250 pp.

CHAMBRE, H. *Union soviétique et développement économique*, Paris (Aubier-Montaigne), 1967, 431 pp.

CHAPMAN, J. G. *Real Wages in Soviet Russia since 1928*, Cambridge (Harvard Univ. Press), 1963, xvi, 395 pp.

DEGRAS, J., NOVE, A. (eds), *Soviet Planning: Essays edited in honour of Naum Jasny*, Oxford 1964, 226 pp.

DOBB, M. *Soviet Economic Development since 1917*, London (Routledge and Kegan Paul), 1966, viii, 315 pp. (6th edn).

ERLICH, A. *The Soviet Industrialisation Debate 1924–1928*, Cambridge (Harvard Univ. Press), 1960, xxiv, 214 pp.

JASNY, N. *The Socialized Agriculture of the USSR*, Stanford (Stanford Univ. Press), 1949, xvi, 837 pp.

JASNY, N. *The Soviet Economy during the Plan Era*, Stanford (Stanford Univ. Press), 1951, xii, 1161 pp.

JASNY, N. *Soviet Industrialisation 1928–1952*, Chicago (Univ. of Chicago Press), 1961, xviii, 467 pp.

KERBLAY, B. *Les marchés paysans en URSS*, Paris (Mouton), 1968, 519 pp.

LAVIGNE, M. *Les économies soviétiques et socialistes*, Paris (Armand Colin), 1979, 415 pp.

LÉWIN, M. *La paysannerie et le pouvoir soviétique 1928–1930*, Paris (Mouton), 1966, 480 pp.

MARCZEWSKI, J. *Planification et croissance économique des démocraties populaires*, Paris, 1956, 2 vols.

MILLAR, J. (ed.) *The Soviet Rural Community*, Urbana, 1970, 416 pp.

NOVE, A. *The Soviet Economy*, London (G. Allen and Unwin), 1968, 375 pp. (3rd edn).

PROKOPOVICZ, S. N. *Narodnoye khozyaystvo SSSR*, New York (Izd. Imeni Chekhova), 1952, 2 vols.

STRAUSS, E. *Soviet Agriculture in Perspective*, London, 1969, 328 pp.

SUCHECKI, W. *Geneza federalizmu radzieckiego* (The origins of Soviet federalism), Warsaw, 1961, 333 pp.

VOLIN, L. *A Century of Russian Agriculture*, Cambridge (Harvard Univ. Press), 1970, 644 pp.

ZALESKI, E. *Planification de la croissance et fluctuations économiques en URSS*, vol. I: *1918–1932*, Paris (Sedes), 1962, xxx, 372 pp.

SOCIETY

ALT, H. and E. *Russia's Children*, New York, 1949, 240 pp.

AZRAEL, J. *Managerial Power and Soviet Politics*, Cambridge (Harvard Univ. Press), 1966.

BAUER, R., WASOILEK, E. *Nine Soviet Portraits*, New York, 1955, 190 p.

BLACK, C. (ed.). *The Transformation of Russian Society*, Cambridge (Harvard Univ. Press), 1967, 695 pp.

CANTRIL, H. *Soviet Leaders and Mastery over Man*, New Brunswick, 1960, xxii, 173 pp.

CONQUEST, R. *Religion in the USSR*, London (Bodley Head), 1968, 135 pp.

DUNHAM, V. *In Stalin's Time. Middle Class Values in Soviet Fiction*, Cambridge Univ. Press, 1976, 283 pp.

FITZPATRICK, S. *The Commissariat of Enlightenment*, Cambridge (Harvard Univ. Press), 1970, 380 pp.

GRANICK, D. *Le chef d'entreprise soviétique*, Paris (L'Entreprise moderne), 1963, 155 pp.

HAYWARD, M., LABEDZ, L. (eds). *Literature and Revolution in Soviet Russia 1917–1962*, London (Oxford Univ. Press), 1963, xx, 235 pp.

INKELES, A. *Public Opinion in Soviet Russia*, Cambridge (Harvard Univ. Press), 1950, xviii, 379 pp.

INKELES, A. *Social Change in Soviet Russia*, Cambridge (Harvard Univ. Press), 1968.

INKELES, A., GEIGER, K. *Soviet Society*, Boston, 1961, 703 pp.

JORAVSKI, D. *Soviet Marxism and Natural Science 1927–1932*, New York (Columbia Univ. Press), 1961, xiv, 433 pp.

KERBLAY, B. *La société soviétique contemporaine*, Paris (Colin), 1977, 304 pp.

KOLARZ, W. *Religion in the Soviet Union*, London (Macmillan), 1961, xii, 518 pp.

KRUGLAK, Th. *The Two Faces of Tass*, Minneapolis (Univ. of Minnesota Press), 1962, 263 pp.

LANE, D. *Politics and Society in USSR*, London, 1972, 616 pp.

LORIMER, F. *The Population of the Soviet Union: History and Prospects*, Geneva, 1946.

SKILLING, H. GORDON, GRIFFITHS, F. (eds). *Interest Groups in Soviet Politics*, Princeton, 1971, 433 pp.

SORLIN, P. *La société soviétique 1917–1967*, Paris (Armand Colin), 1967, 281 pp. (2nd edn).

STRUVE, G. *Soviet Russian Literature 1917–1950*, Norman (Univ. of Oklahoma Press), 1951, 414 pp.

STRUVE, N. *Les chrétiens en URSS*, Paris (edn du Seuil), 1963, 374 pp.

ZALESKI E., et AL., *La politique de la science en URSS*, Paris (OCED), 1969, 637 pp.

Journals of which particularly extensive use has been made:
Proletarskaya Revolyutsiya
Krasnyy Arkhiv
Krasnaya Letopis'
Voprosy Istorii
Voprosy Istorii KPSS (published since 1937)
Voprosy Filosofii
Partiynaya zhizn'
Bol'shevik, then *Kommunist*
Pravda and *Izvestiya* (and from 1949–50 the local editions of *Pravda*).
Spravochnik Partiynogo Rabotnika (rather irregular annual collection from 1921 to 1936, containing decrees and other Party documents).

BIBLIOGRAPHICAL NOTES

In order not to overload this book, footnotes have been omitted. The bibliography which indicates mainly the Western sources is far from being exhaustive and does not include all the works which have been consulted. As a general rule, each time that original sources exist and were accessible they were used and completed by secondary, but reliable works. The notes which follow indicate some particular sources, used in the different chapters. The most frequently consulted periodicals are shown in the general bibliography.

CHAPTER ONE

The police apparatus

DALLIN, A., BRESLAUER, G. *Political Terror in Communist Systems*, Stanford Univ. Press, 1970.

DZHERZHINSKY, F. *Izbrannye proizvedeniya v dvukh tomakh* (Selected Works), Moscow, 1957, 2 vols.

GLADKOV, T., SMIRNOV, M. *Menzhinsky*, Moscow, 1969, 366 pp.

JASNY, N. 'Labour and output in Soviet concentration camps', *Journal of Political Economy*, 59, No. 5, Oct. 1951, 405–19.

LATSIS, M. *Chrezvychainye Kommissii po bor'be s kontrrevolyutsyey* (The Extraordinary Commissions for the Fight against the Counter-Revolution), Moscow, 1924.

MIRONOV, N. R. *Programma KPSS: voprosy dal'neyshego ukrepleniya zakonnosti i pravoporyadka*, Moscow, 1962, pp. 6 ff.

SOLZHENITSYN, A. *Arkhipelag Gulag, 1918–1956: opyt khudozhestvennogo issledovaniya*, Paris, YMCA, 3 vols, 1973–75.

CHAPTER TWO

Collectivisation and industrialisation

ABRAMOV, B.A. 'Likvidatsiya kulachestva kak klass na osnove sploshnoy kollektivizatsii sel'skogo khozyaystva', *Istoricheskiy Zhurnal*, **38**, 1951. 3–47.

AVDEYEV, V. K. *'Likvidatsiya Kulachestva v nizhne-volzhskom kraye'*, *Istoriya SSSR*, **1**, 1958.

BOGACHEV, 'Rost rabochego klassa SSSR v gody pervoy pyatiletki' (The growth of the working class in the USSR during the years of the First Five-Year Plan), *Voprosy Istorii*, **8**, 1953.

DOBB, M. *Soviet Economic Development since 1917*, London, 1947.

GUGUSHYILI, P. V. (ed.), et al. *Istoriya kollektivizatsii sel'skogo khoziyastva gruzinskoy SSSR (1927–37 gg)*. (History of the collectivisation of agriculture in Georgia (1927–37), Leningrad, 1970, 771 pp.

Istoriya Kolkhoznogo prava, sbornik zakonodatel'nykh materialov SSSR i RSFSR 1917–58 (History of kolkhoz law) Moscow, 518 and 659 pp.

Istoriya SSSR s drevneyshikh vremen do nashikh dney, Vol. VIII: *Bor'ba sovetskogo naroda za postroyeniya fondamenta sotsializma v SSSR, 1921–32* (The fight of the Soviet people for the building of the foundations of socialism in the USSR, 1921–32), Moscow, 1966, 728 pp.

IVANOV, L. M. (ed.). *Rossiiskiy proletariat: oblik, bor'ba, gegemoniya* (The Russian proletariat, physiognomy, struggle, hegemony), Moscow, 1970, 364 pp.

KHAVIN, A. F. 'Iz Istorii promyshlennogo stroitel'stva na vostoke' (History of the construction of industry in the east), *Voprosy Istorii*, **5**, 1960.

MARKOV, S. F. 'Ukrepleniye sel'skikh partiynykh organizatsiy v period podgotovki massovogo kolkhoznogo dvizhenya' (The strengthening of the rural organisations of the Party during the period of the kolkhoz mass movement), *Voprosy Istorii KPSS*, **3**, 1962.

MEQUET, G. 'Autour du premier plan quinquennal', *Annales d'Histoire économique et sociale*, May 1932, pp. 257–94.

NAZNANOV, S. V. (ed.). *Partiya, Organizator Kolkhoznogo stroya* (The Party, the organiser of the kolkhoz structures), Moscow, 1958.

Ocherki istorii kollektivizatsii sel'skogo khozyaystva v soiuznykh respublikakh (Outlines of the collectivisation of agriculture in the federated republics), Moscow, 1963.

PAVLOV, I. V. *Kolkhoznoye pravo*, Moscow, 1960, 373 pp.

POSPELOV, P. N. (ed.) et al. *Leninskiy plan sotsialisticheskoy industrializatsii i ego osushchestvleniye* (The Leninist plan of socialist industrialisation and its practical application), Moscow, 1960, 384 pp.

PROKOPOVICH, S. N. *Narodnoye Khozyaystvo SSSR*, vol. I.

ROMANOVA, Z. G. *Deyatel'nost' kommunisticsheskoy partii Moldavii po razvitiyu promyshlennosti respubliki (1924–65 gg.*) (The activity of the Moldavian Communist Party in the development of the industry of the republic (1924–65), Kishinev, 1970.

SELEZNEV, V. A., STARIKOVA, A. Ja. (eds). *Kollektivizatsiya sel'skogo khozyaystva v Severo-Zapadnom rayone (1927–37 gg)* (The collectivisation of agriculture in the north-west region (1927–37), Leningrad, 1970, 424 pp.

VAKHABOV, M. G., MALYKHIN, F. G. (eds). *Osushchestvleniye leninskikh idey industrializatsii v Uzbekistane* (The implementation of Leninist ideas on industrialisation in Uzbekistan), Tashkent, 1970, 408 pp.

CHAPTER THREE

ALEXANDROV, P. 'Priem v nashu partiyu luchikh lyudey nashey rodiny' *Partiynoye stroitel'stvo*, 1, 1938.

AVTORKHANOV, A. *Stalin and the Soviet Communist Party: A study in the technology of power*, New York, 1959, Chs. 27–28.

BAKSHIYIEV, D. *Partiynoye stroitel'stvo v usloviyakh pobedy sotsializma v SSSR* (The construction of the Party in the conditions of the victory of Socialism in the USSR), Moscow, 1954.

BARMINE, A. *One who Survived: The life of a Russian under the Soviets, New York, 1951.*

BARSUKOV, N., YUDIN, I. 'Razshireniye sotsial'noy bazy KPSS' (The enlargement of the social base of the CPSU), *Politicheskoye samoobrazovaniye*, 15, 1935.

BECK, F., GODIN, W. *Russian Purge and the Extraction of Confession*, New York, 1951.

BERIYA, L. *K voprosu ob istorii bol'shevistkikh organizatsiy v zakavkaze*, Moscow, 1952.

BERZUKOVA, *Kommunisticheskaya partiya Uzbekistana v tsifrakh sbornik statisticheskikh materialov 1924–64 gody* (The CP of Uzbekistan in figures, collection of statistical documents, 1924–64), Tashkent, 1964, 212 pp.

DAROCHECHE, B. 'Mykola Skrypnyk et la politique d'ukrainisation', *Cahiers du monde russe et soviètique*, 1–2, 1971.

DENKO, H. *The Modern Inquisition*, London, 1953.

DEWAR, H. 'Murder revisited: The case of Sergey Mironovich Kirov', *Problems of Communism*, Sept.–Oct., 1965.

EHRENBURG, I. *Memoirs 1921–41*, New York, 1964.

FEUCHTWANGER, L. *Moscow 1937: My visit described for my friends*, London, 1937.

FRENKEL, A. 'O predstoyashchey chistki partii' (On the future purge of the Party), *Partiynoye stroitel'stvo*, Nos. 23–24, 1932.

GARTHOFF, R. L. *How Russia Makes War*, London, 1954.

HELLER, M. *Le monde concentrationnaire et la littérature soviétique*, L'age d'Homme, 1974, 317 pp.

KAZIEV, M. *Nariman Narimanov, Zhizn' i deyatel'nost'* (Nariman Narimanov, life and activity), Baku, 1970, 188 pp.

KELENDZHERIDZE, A. K. *Sergo Ordzhonikidze zhurnalist*, Tbilisi, 1969, 88 pp.

KHODZHAYEV, F. *Izbrannyye trudy v trekh tomakh*, Tashkent, 1970, vol. I.

LERMOLO, E. *Face of a Victim*, New York, 1937, *Letter of an old Bolshevik*, New York, 1937.

MANDELSTAMM, N. *Contre tout espoir*, Paris, Gallimard, 1971–75, 3 vols. I and II, 304 pp.; III, 330 pp.

ORLOV, A. *The Secret of Stalin's Crimes*, New York, 1953.

POSTYSHEV, P. 'Tekushchiye zadachi marksistsko-leninskogo vospitaniya (The current tasks of Marxist-Leninist education), *Partiynoye stroitel'stvo*, 15–16, 1931.

'Proverka partdokumentov, ser'eznoe ispytaniye partiynykh kadrov' (The verification of the documents of the Party, a serious test for the Party cadres), *Partiynoye stroitel'stvo*, 15, 1935.

RUMYANTSEV, I. 'Povtornaya proverka partdokumentov' (The principal lessons of the verification of Party documents), *Partiynoye stroitel'stvo*, 2, 1936.

SHIVIREV, F. 'K voprosu o rabote s kommunstami odinochkami' (On work with isolated communists), *Partiynoye stroitel'stvo*, 14, 1935.

SOLZHENITSYN, A. *Arkhipelag Gulag, 1918–56: opyt, khudozhest-*

vennogo issledovaniya, Paris, YMCA, 3 rols, 1973–75 (*The Gulag Archipelago*, Fontana, London, 1976–8).

The Case of Leon Trotsky; Report of hearings before the Dewey Commission, New York, 1937.

VLASOV, V. 'Nedostatki komsomola v ukreplenii ryadov partii' (The inadequacies of the Komsomol in the strengthening of the ranks of the Party), *Partiynoye stroitel'stov*, **21**, 1930; 'O kurse priema novykh chlenov v VKP (b)' (On the admission of new members into the CPSU), *Partiynoye stroitel'stvo*, **15**, 1938.

WEISSBURG, A. *L'accusé*, Paris, 1953.

YAROSLAVSKY, E. *Za bol'shevitskuyu proverku i chistku ryadov* (For a Bolshevik verification and purge in our ranks), Moscow, 1933.

CHAPTER FOUR

BLACK, C. E. (ed.). *Rewriting Russian History*, New York, 1956.

Gosudarstvennyy plan razvitiya Narodnogo Khozyaystva SSSR na 1941 god. (The State Plan for the development of the economy for the year 1941), Baltimore.

Diplomaticheskiy slovar' (Diplomatic dictionary), vol. I, Moscow, 1948; vol. II, Moscow, 1950 and vol. III, 1960–64.

Entsiklopediya gosudarstva i prava (Encyclopedia of State and law), Moscow, 1925–28, 3 vols.

HODGMAN, D. R. *Soviet Industrial Production, 1928–51*, Cambridge (Mass.), 1954.

LEPESHKIN, A. L. *Sovety vlast' naroda* (The Soviets, popular power), vol. I, 1917–36, Moscow, 1966; vol. II, 1939–67, Moscow, 1967.

MAZOUR, A. G. *Modern Russia Historiography*, Princeton, 1958, 2nd edn.

NECHKINA, M. V. (ed.). *Istoriya SSSR* (History of the USSR), vol. II, *Rossiya v XIX veke* (Russia in the nineteenth century), Moscow, 1940.

PANKRATOVA, A. M. 'Velikiy russkiy narod i ego rol' v istorii', (The Great Russian people and its role in history), *Prepodavaniye Istoriya v shkole*, **5**, 1946.

PANKRATOVA, A. M. *Druzhba narodov SSR, osnova osnov mnogonatsional'nogo sotsialisticheskogo gosudarstva* (The friendship of the peoples of the USSR, foundation of the foundations of a plurinational socialist State), Moscow, 1953.

Stalin

PASHUKANIS, E. B. *Obshchaya teoriya prava i Marksism* (The general theory of law and Marxism), Moscow, 1928, 4th edn.

POKROVSKY, M. N. *Diplomatiya i voyny tsarskoy Rossii v XIX stoletii* (The diplomacy and wars of Tsarist Russia in the nineteenth century), Moscow, 1923.

POKROVSKY, M. N. *Istoricheskaya nauka i bor'ba klassov* (Historical science and the class war), Moscow, 1933, vol. I.

POPOV, A. L. 'Iz istorii zavoyevaniya sredney Azii' (On the history of the conquest of Central Asia), *Istoricheskiye zapiski*, **IX**, 1940.

PROKOPOVICH, S. N. *Narodnoye Khozyaystvo SSSR*, vols I and 2, pp. 120 ff.

Protiv antimarksistskoy kontseptsii M. N. Pokrovsky sbornik statey (Against the anti-Marxist ideas of Pokrovsky, collection of articles), Moscow, 1940, especially the article by A. L. Popov.

Protiv istoricheskoy kontseptsii M. N. Pokrovskogo (Against the historical ideas of M. N. Pokrovsky), Moscow, 1939, vol. I.

PUNDEFF, M. (ed.). *History in the USSR: Selected readings*, Stanford, 1967.

SHESTAKOV, A. V. *Kratkiy kurs istorii SSSR* (Short course of the history of the USSR), Moscow, 1937.

STUCHKA, P. *Leninizm i gosudarstvo* (Leninism and the State), Moscow, 1925.

Teoriya gosudarstva i prava (The theory of State and law), Moscow, 1949.

Trudy pervoy vsesoyuznoy konferentsii istorikov-marksistov (Proceedings of the first All Union Conference of Marxist Historians), Moscow, 1930.

Verkhovnyy Soviet SSR (The Supreme Soviet of the USSR), Moscow, 1967.

VYLTSAN, M. A. *Sovetskaya derevnya nakanune Velikoy Otechestvennoy voyny, 1938–41 gg.* (The Soviet village on the eve of the Great Patriotic War 1938–41), Moscow, 1970, 200 pp.

VYSHINSKY, A. *Voprosy teorii gosudarstva i prava* (Problems of the theory of State and law), Moscow, 1949.

Society

GUDOV, I. *Put' Stakhanovtsa, Rasskaz o moey zhizni* (The way of a stakhanovite), Moscow, 1938.

SMIRNOV, F. *Pavlik Morozov v pomoshch pionervozhatomu* (Pavlik Morozov: To help the responsible pioneer), Moscow, 1938.

'Literatura Stakhanovskogo dvizheniya, *Literaturnaya Gazeta*, 29 Oct. 1935, No. 60.

CHAPTER FIVE

On German policy in the USSR and the Soviet military reaction

DEBORIN, A. *O kharaktere vtoroy mirovoy voyny* (On the character of the Second World War), Moscow, 1960.

DEBORIN, A., TELPUKHOVSKY, B. *Itogi i uroki vtoroy mirovoy voyny*, Moscow, 1970, 344 pp.

Documents des archives du centre de documentation juive contemporaine, Paris, series CXL.

FIRESIDE, H. *Icon and Swastika, the Russian Orthodox Church under Nazi and Soviet Control*, Cambridge (Mass.), 1971, 245 pp.

FISCHER, G. *Soviet Opposition to Stalin*, Cambridge (Mass.), 1952.

GOEBBELS, J. *Diaries, 1942–43*, New York, 1948.

GRIGORIENKO, P. *Staline et la Deuxième Guerre Mondiale*, Paris, 1969, pp. 39–142.

Hitler's Table Talk, 1941–44, London, 1953.

NIKOLAY (Metropolitan), *Slova, rechi, posylaniya, 1941–46* (Words, speeches and addresses), Moscow Patriarchy, 1947.

Nyurenbergskiy protses – sbornik materialov i dokumentov (The Nuremberg Trials; Collection of materials and documents), Moscow, 1954, 2 vols, 936 and 1150 pp.

RAUSCHNING, H. *Hitler's Aims in War and Peace*, London, 1940.

STEENBERG, S. *Vlassov, verrater oder Patriot*, Cologne, 1968.

STRICK-STRIKFELDT, W. *Contre Stalin et Hitler*, Paris, 1971, 253 pp.

TELPUKHOVSKY, B. S. *Velikaya otechestvennaya voyna sovetskogo soyuza 1941–45* (The Great Patriotic War of the Soviet Union), Moscow, 1959.

Trial of the Major War Criminals, 42 vols, Nuremberg, 1947–49.

Vneshnyaya politika sovetskogo soyuza v period otechestvennoy voyny (The foreign policy of the Soviet Union during the Great Patriotic War), Moscow, 1946–47, 3 vols.

On the partisans

HESSE, E. *Der Sowjet russische partisanen Krieg*, Göttingen, 1969, 292 pp.

Sovetskiye partizany, Moscow, 1961, 830 pp.

Stalin

On the economic policy of the USSR during the war

ARUTYUNYAN, ZHU. V. *Sovetskoye krestianstvo v gody velikoy otechestvennoy voyny (1941–1945)* (The Soviet peasantry during the Great Patriotic War), Moscow, 1953, 459 pp.

BELONOSOV, T. *Sovetskiye profsoyuzy v gody voyny* (The Soviet trades unions during the war), Moscow, 1970, 216 pp.

BORKOVSKY, D. A. 'Vostanovleniye i organizatsionnoye ukrepleniye partiynykh organizatsii zapadnykh oblastey 1944–45' (The reconstruction and organisational reinforcement of Party organisations in the western Ukraine), *Voprosy Istorii*, **II**, 1971, pp. 65–74.

CHERNYAVSKY, V. G. *Voyna i prodovolstviye, snabzheniya gorodskogo naseleniya v otechestvennuyu voynu* (War and trade) Moscow, 1964, 208 pp.

Eshelony idut na vostok, Moscow, 1966, 263 pp.

GLADKOV, I. (ed.). *Sovetskaya ekonomika v period velikoy otechestvennoy voyny (1941–45)* (The Soviet economy during the Great Patriotic War), Moscow, 1970, 504 pp.

Ideologicheskaya rabota KPSS na fronte 1941–45 (The ideological work of the CPSU at the front), Moscow, 1960.

Eshelony idut na vostok, Moscow, 1966, 263 pp.

Bibliographical articles on the Second World War by KARESEV, A. V., *Voprosy Istorii*, 6, 1961; GROZDEV, I. I., *Voprosy Istorii KPSS*, 5, 1970; KUKIN, D. M., *Voprosy Istorii*, *KPSS*, 6, 1971.

Istoriya uzbekskoy SSSR, vol. II, Tashkent, 1957, pp. 590 ff.

KALININ, M. *Voprosy sovetskogo stroitel'stva: stat'i i rechi*, Moscow, 1958, pp. 670 ff.

KUMANEV, G. A. *Sovetskiye zheleznodorozhniki v gody velikoy otechestvennoy voyny (1941–45)* (The Soviet railwaymen during the Great Patriotic War), Moscow, 1963, pp. 61 ff.

MITROFANOVA, A. V. *Rabochiy klass sovetskogo soyuza v pervyy period velikoy otechestvennoy voyny* (The working class of the USSR in the first period of the Great Patriotic War), Moscow, 1960, pp. 90 ff.

MOREKHINA, G. 'Vostanovleniye narodnogo khozyaystva na osvobozhdennoy territorii v period velikoy otechestvennoy voyny', *Voprosy Istorii*, 8, 1961, pp. 41–60.

Revue d'Histoire de la deuxième guerre mondiale, special number, *L'URSS en guerre*, No. 43, July, 1961.

SHIGALIN, G. *Narodnoye Khozyaystvo SSSR v period velikoy*

otechestvennoy voyny (Soviet agriculture during the Great Patriotic War), Moscow, 1960, 277 pp.

VASIL'YEV, A. F. 'Deyatel'nost' partiynykh organizatsii yuzhnogo urala po razmeshcheniyu evakuirovanykh predpriatii v 1941–42 g' (The activity of the organisations of the Party of the south Urals in the setting up of evacuated industries, 1941–42), *Voprosy Istorii*, **6**, 1961, pp. 63–70.

Voprosy Istorii KPSS perioda velikoy otechestvennoy voyny (Problems of the history of the CPSU in the period of the Great Patriotic War), Kiev, 1961.

Voprosy truda v SSSR, Moscow, 1958, pp. 30 ff.

VOZNESENSKY, N. *Voyennaya ekonomika SSSR v period otechestvennoy voyny*, (The economy of the USSR during the Great Patriotic War), Moscow, 1947, pp. 70 ff.

Chapters on the war in:

Ocherki istorii kommunisticheskoy partii Turkmenistana, Ashkhabad, 1961: *Ocherki istorii kommunisticheskoy partii Uzbekistana*, Tashkent, 1964; *Ocherki istorii kommunisticheskoy partii Tadzhikistana*, Dushambe, 1964; *Ocherki istorii kommunisticheskoy partii Kirgizii*, Frunze, 1966.

Sovetskaya zhenshchina v velikoy otechestvennoy voyne (Soviet women during the Great Patriotic War), Alma Ata, 1942.

CHAPTER SIX

Industrial reconstruction

KERBLAY, B. 'L'évolution de l'alimentation rurale en Russie, 1896–1960', *Annales*, 5, 1962, 885–913.

BERGSON, A. (ed.). *Soviet Economic Growth, Conditions and Perspectives*, Columbia, 1953, 376 pp.

Ekonomika SSSR v poslevoyennoy period (The Soviet economy in the post-war period), Moscow, 1962, 488 pp.

GRIGOROFF, G. 'La population et le logement en URSS', *Population*, June 1969, 77–8.

JASNY, N. *Soviet Prices of Producers' Goods*, Stanford, 1952.

MOORSTEIN, R. *Prices and Production of Machinery in the Soviet Union*, Harvard, 1962, 498 pp.

Narodnoye khozyaystvo SSSR, Statisticheskiy sbornik (The Soviet economy: Collection of statistics), Moscow, 1956.

Stalin

Promyshlennost' SSSR, Statisticheskiy sbornik (Soviet industry: Collection of statistics), Moscow, 1957.

SENYAVSKY, S. L. TEL'PUKHOVSKY, V. B. *Rabochiy klass SSSR, 1938–56* (The Soviet working class), Moscow, 1971, 536 pp.

Reconstruction of agriculture

KERBLAY, B. 'L'évolution de l'alimentation rurale en Russie, 1896–1960', *Annales*, 5, 1962, 885–913.

LAIRD, R. 'The new zveno controversy', *Osteuropa Wirtschaft*, 4, 1966. 254–61 and *Economie soviétique et économies planifiées*, No. 8, March 1970, especially p. 9.

SCHILLER, S. *Die Landwirtschaft der Soviet Union 1917–53*, Tübingen, 1954, pp. 91 ff.

STALIN, J. V. *The Economic Problems of Socialism in the USSR*,

ZELENIN, I. *Sovkhozy SSSR 1941–1950 gg.* (The sovkhozes of the USSR), Moscow, 1969, 344 pp.

For the controversy on the agrocities: L. Richter's contribution in *Soviet Planning: Essays in honour of Naum Jasny*, ed. by Jane Degras and Alec Nove, Oxford, 1964, 225 pp. and *Kommunist*, 2, No. 21, 1951 (criticism of Aryutyunov); *Bolshevik*, No. 18, Sept. 1952.

Communist integration of the countryside

AFANASYEV, L. *'Ob ukrepleniye pervychnykh partiynykh organizatsii v kolkhozakh'*, *Partiynoye stroitel'stvo*. Nos. 7, 8, April 1946, 22.

KHRUSHCHEV, N. S. *O merakh dal'neyshego razvitiya sel'skogo khozyaystva SSSR* (The methods for the ulterior development of agriculture in the USSR), Moscow, 1959, pp. 72 ff; *Stroitel'stvo kommunizma v SSR i razvitiye sel'skogo khozyaystva*, Moscow, 1962–64, 8 vols. (The building of communism in the USSR and the development of agriculture.)

CHAPTER SEVEN

The nationalities

BAGIROV, M. D. 'K voprosu o kharaktere dvizheniya myuridizma i Shamilya' (On the character of the Murid movement and of Shamil), *Bolshevik*, 13, 1950, 21–37.

BARGHOORN, F. 'Stalinism and the Russian cultural heritage', *Review of Politics*, **XIV**, 1952, 178–203.

ERZHANOV, A. E. *Uspekhi natsional'noy politiki KPSS v Kazakhstane* (1946–58) (The successes of the national policy of the CPSU in Kazakhstan), Alma Ata, 1969, 251 pp.

KHRUSHCHEV, N. S. 'Stalinskaya druzhba narodov, zalog nepobedimosti nashey rodiny' (Stalinist friendship of the peoples, sign of our Party's invincibility), *Bolshevik*, **24**, 1949, 80–5.

LITVIN, 'Ob istorii Ukkrainskogo naroda' (On the history of the Ukrainian people), *Bolshevik*, **7**, 1947, 41–56; 'O zadachakh sovetskikh istorikov v borbe s proyavleniyem burzhuaznoy ideologii' (On the tasks of Soviet historians in the fight against bourgeois ideology), *Voprosy Istorii*, **2**, 1949, 3–13.

MATYUSHKIN, N. L. *Sovetskiy patriotizm moguchaya dvizhushchaya sila sotsialisticheskogo obshchestva* (Soviet patriotism: A powerful and dynamic force in Soviet society), Moscow, 1951.

PANKRATOVA, A. M. *Velikiy russkiy narod* (The Great Russian people), Moscow, 1952; 'Zadachi sovetskikh istorikov v oblasti novoy i noveyshey istorii' (The tasks of Soviet historians in the field of recent history), *Voprosy Istorii*, **3**, 1949, 3–13.

TOLSTOY, S. P. 'Velikaya pobeda leninsko-stalinskoy natsional'-noy politiki' (The great victory of the Leninist-Stalinist national policy), *Sovietskaya etnografiya*, **I**, 1950, 3–23.

The ideological controversy

For this part, I have principally used the journals: *Voprosy Filosofii*, *Voprosy Ekonomiki*, *Mirovoye Khozyaystvo i Mirovaya Politika* and *Literaturnaya gazetta*.

DANIELS, R., 'State and Revolution: a case study in the genesis and transformation of communist ideology', *American Slavic and East European Review*, **XII**, No. I, 1953, 22–43.

KAISER, M. 'Le débat sur la loi de la valeur en URSS, étude rétrospective, 1941–53', *Annuaire de l'URSS*, 1965, 555–69.

MEDVEDEV, J. Grandeur et chute de Lysenko, Paris, 1971, 306 pp.

O sovetskom sotsialisticheskom obshchestve, Moscow, 1948, 566 pp.

Report on the scientific session devoted to Stalin's work in *Voprosy Filosofii*, **3**, 1952, 240–261.

STALIN, J. V. *Marksism i voprosy yazykoznaniya* (Marxism and linguistic problems), Moscow, 1952.

VARGA, E. *Ocherki po probleme politekonomiki kapitalizma* (Essay on

the problems of capitalist economic policy), Moscow, 1965,
pp. 42–51.

ZHDANOV, *The Strategy and Tactics of International Communism*,
Moscow, 1947.

On the crisis of capitalism, see also the articles in *Voprosy ekonomiki*
on G. V. Kozlov, No. 4, 1952, pp. 68 ff; by M. Rubinstein, in
No. 10, 1952, pp. 38–55; by Trakhtenberg in No. 10, 1952, 69–
85; by I. Lemin in No. 12, 1952, 34–53; by E. Varga in
No. 4, 1950, 48–71, in which Varga deals especially with the
decline of British imperialism.

CHAPTER EIGHT

Party life after the war

BUGAYEV, E., LEIBZON, B. *Besedy ob ustave KPSS* (Talks on the
Party statues) Moscow, 1964, pp. 35 ff.

GALKIN, K. T. *Vyssheye obrazovaniye i podgotovka nauchnykh kadrov
v SSSR* (Higher education and the formation of cadres in the
USSR), Moscow, 1958, pp. 161 ff.

KALININ, V. B., NECHIPURENKO, V. I., SAVEL'YEV, V. M. *Kommunisticheskaya partiya v velikoy otechestvennoy voyne; dokumenty i
materialy* (The Party during the war; materials and documents),
Moscow, 1970, 496 pp.

KOZLOV, F. 'Politicheskaya bditel'nost', ob'yazanost' chlena partii'
(Political vigilance, the duty of the Party member), *Kommunist*,
1953, 46–58.

KULTUCHEV, S. 'Rost ryadov partii v 1945–1950 gg.', *Voprosy
Istorii KPSS*, 2, 1958, 60 ff; 'Voprosy chlenstva VKP(b)', *Partiynaya Zhizn'*, Oct. 1947, 82.

Kulturnoye stroitel'stvo SSSR: statisticheskiy sbornik (Cultural development in the USSR: statistical collection), Moscow, 1956,
pp. 248 ff.

Problems of power before 1953

ARMSTRONG, J. *The Politics of Totalitarianism*, New York, 1961,
Chs. XV, XVI, XVII.

AVTORKHANOV, A. *The Communist Party Apparatus*, Chicago,
1966, 422 pp.

Bibliographical notes

NIKOLAYEVSKY, B. 'Rastrel Ryumina', (The execution of Ryumin), *Sotsialisticheskiy Vestnik*, 8, Sept. 1954.

The post-war purges

For the Caucasus, see *Zarya Vostoka* and *Les crises du parti communist de la RSS de la Géorgie*, Chroniques étrangères, URSS, Paris, documentation française, No. 162, 1956, 13–22.

BERZUKOVA, *Kommunisticheskaya partiya Uzbekistana v tsifrakh* (The CP of Uzbekistan in figures), Tashkent, 1964, 212 pp.

VATOLINA, L. 'Izrail, baza amerikanskogo imperializma na blizhnem vostoke' (Israel, the base of American imperialism in the Near East) *Voprosy Istorii*, 4, 1951, 94–105.

CHAPTER NINE

ALLILUYEVA, S. the two books quoted in the general bibliography.

Bor'ba KPSS za zaversheniye stroitel'stva sotsializma 1953–58 (The fight of the CPSU for the achievement of the building of socialism), Moscow, 1961, pp. 65–195.

BREZHNEV, L. I. *Voprosy agrarnoy politiki KPSS rechi i doklady* (Problems of agricultural policy, speeches and reports), Moscow, 1974.

BULGANIN, N. A., KHRUSHCHEV, N. S. *Rechi vo vremiya prebyvaniya v Indii, Birme i Afganistane*: Noyabr'- dekabr' 1955 (Speeches given in India, Burma and Afghanistan), Moscow, 1955.

Istoriya KPSS (On the XXth Congress and its problems), Moscow, 1962, pp. 650 ff.

KHRUSHCHEV, N. S. *Stroitel'stvo Kommunizma v SSSR i razvitiye sel'skogo khozyaystva* (The building of commmunism in the USSR and the development of agriculture), Moscow, 1962, 3 vols. This collection is usually called the Agricultural Speeches.

Khrushchev Remembers, Little, Brown and Co., Boston, 1970. The secret report is included in pp. 559–618. In spite of doubts about its authenticity this work is very useful.

MEDVEDEV, R. *Le Stalinisme*, Seuil, Paris, 1972, 638 pp.

MEDVEDEV, R. and KHRUSHCHEV, J. Maspero, Paris, 1977.

NEKRICH, A. *Otreshis' ot strakha*, London, 1979, 414 pp.

Spravochnik partiynogo rabotnika (Diary of the Party worker), Moscow, 1957, 611 pp.

Stalin

XX s'ezd Kommunisticheskoy Partii Sovetskogo Soyuza, Steno-graficheskiy Ochet (The XXth Congress of the CPSU), Moscow, 1956–58, 2 vols, I, 640 pp. and II, 560 pp.

XXII s'ezd Kommunisticheskoy Partii Sovetskogo Soyuza (XXII Congress of the CPSU), Moscow, 1962, 3 vols (vol. I pp. 582 ff. deals with the problems of the XXth Congress).

Vsesoyuznoye soveshchaniye istorikov (Conference of the historians of the USSR on the problems of the past), Moscow, 1964.

In addition to the above works this chapter relies above all on the systematic analyses of *Pravda, Izvestia, Literaturnaya Gazeta, Voprosy Istorii* and *Kommunist* in 1953–56.

INDEX

Index

Index

Index

Nevsky, A., 62–3, 110, 112
newspapers *see* journals
Nikolayev, 32–4
NKVD (People's Commissariat of
    Internal Affairs), 11, 32, 34, 43
Nogin, V. P., 223, 225
non-Russians *see* nationalities
Nuzratallah, M., 46

OGPU (Unified Political Administration
    of State), 7–13, 18
'Old Guard" *see* Bolsheviks
Ordzhonikidze, 40, 153, 225, 230
Orgburo, 225–30
Ostpolitik of Germany, 89–108
Ozenbashly, A., 104

Pamfilov, K. D., 86
Pankratova, A. M., 152, 251, 257
parents, 76; *see also* family
Partiinaya Zhiza, 158
Partiinoe Stroitelstvo, 37
Partiinost, 3, 159
Pashukanis, E. B., 36, 56–7, 252
passports, internal, 24–5, 217
Patolichev, N. S., 231
patriotism, 110–13; *see also* nationalism
Pauker, A., 180
Pavlov, I. V., 110, 249
peace, 186; post-war, 134–69
peasantry, 3–4: and Communist Party,
    140–5; and free enterprise,
    71–2, 121, 140, 142–5; as target
    for modernisation, 168, 216–17,
    220; *see also* agriculture; working
    class
Pechkov, 41
Pegov, N. N., 231
personality cult, 124–6, 204–7,
    218–20; *see also* heroes
Pervukhin, M. G., 86, 230–1
Peter the Great, 62–3, 66
Petrovsky, G. I., 224–5, 228–30
Piatakov, Y. L., 39–40, 234
Pikel, 39
Platonov, 159
Plekhanov, 110
Pokrovsky, M. N., 61–5, 252
Poland, 4, 92, 111, 119, 126, 132,
    180
police, 6–14, 217

Politburo, 171–3, 224–30
political conditions, improved, 196–7;
    see also government
political conflict, 175–9
Ponomarenko, P. K., 230–1
Popkov, P., 86, 176
population, 19–21, 44; evacuation,
    86–8, 98; *see also* deportations;
    repression
Poskrebyshev, 171
Pospelov, P. N., 231, 249
post-Stalinism, 189–222
post-war reconstruction, 134–69, 176,
    179–85
Postyshev, P. P., 229–30, 250
prison *see* camps
power: industrial, 22; political, 1–14,
    212–15
Pozharski, D., 110
Pravda, 246, 260; on agriculture, 18,
    143–4, 146; on Communist
    Party, 5; on economics, 163; on
    Jews, 119; on NKVD, 33–4; on
    Skrypnik, 46; on Stalin, 202; in
    wartime, 84; on Yugoslavia, 198;
    see also journals
pravoslaviye, 215
Preobrazhensky, E., 224, 234
Presidium, 187, 230–1
press *see* journals
prices, 135–6, 139–40, 195
Prokofiev, S., 159
proletariat, 22–3, 83, 137; *see also*
    peasantry
Proletarskaya Revoliutsya, 61, 246
propaganda, German, 107–8
Propagandist, 113
Pugachev, E., 62
purges: army, 48–50, 88–9; of
    Bolsheviks, 44, 50–2, 171; in
    Communist Party, 28–30, 35–8,
    44–5, 50–2, 201; denounced,
    196–7; of 1930s, 3, 28–30,
    45–52; in NKVD, 34; post-war,
    176, 179–85; religious, 116; *see
    also* repression
Pushkin, A., 110
Puzanov, A. M., 231
Pyatnitsky, I. A., 225

Qari, M., 46

266

Index